Handbook for the Crisis Communication Center

This book is part of the Peter Lang Media and Communication list.
Every volume is peer reviewed and meets
the highest quality standards for content and production.

PETER LANG
New York • Bern • Frankfurt • Berlin
Brussels • Vienna • Oxford • Warsaw

Bolanle A. Olaniran and Juliann C. Scholl

Handbook for the Crisis Communication Center

PETER LANG
New York • Bern • Frankfurt • Berlin
Brussels • Vienna • Oxford • Warsaw

Library of Congress Cataloging-in-Publication Data

Names: Olaniran, Bolanle Abodunrin, author. | Scholl, Juliann C., author.
Title: Handbook for the crisis communication center /
Bolanle A. Olaniran, Juliann C. Scholl.
Description: New York: Peter Lang, 2016.
Includes bibliographical references and index.
Identifiers: LCCN 2015042703 | ISBN 978-1-4331-2434-1 (hardcover: alk. paper)
ISBN 978-1-4331-2433-4 (paperback: alk. paper) | ISBN 978-1-4539-1759-6 (ebook pdf)
ISBN 978-1-4331-3563-7 (epub) | ISBN 978-1-4331-3564-4 (mobi)
Subjects: LCSH: Emergency management. | Communication in management.
Classification: LCC HV551.2.O43 2016 | DDC 363.34068/4—dc23
LC record available at http://lccn.loc.gov/2015042703

Bibliographic information published by **Die Deutsche Nationalbibliothek.**
Die Deutsche Nationalbibliothek lists this publication in the "Deutsche Nationalbibliografie"; detailed bibliographic data are available on the Internet at http://dnb.d-nb.de/.

29 Broadway, 18th floor, New York, NY 10006
www.peterlang.com

CONTENTS

FOREWORD

H. Dan O'Hair
University of Kentucky

Prolonged ice storms, tornadoes, pandemic flu, violence in schools, floods, draughts, high-impact transportation crashes, and wild fires, these are just a few of the poignant reminders that risks and crises are ever ready to impose their presence in our communities. Risks, crises, and disasters are locally situated—they occur in communities. While community members fully expect to be protected against threats, they find themselves more vulnerable today than ever before. Managing information about risks, crises, and disasters poses overwhelming challenges for individuals, for families, and for communities.

Amid many economic and political changes, homeland security strategists have the responsibility and opportunity to chart directions for preparedness and for responses that transcend political challenges and that elevate community safety to new levels. Many organizations stake claim to the onerous task of securing our communities and of promoting the health and well-being of our citizens. The multitude of state, county, municipal, nonprofit, and faith-based organizations deliver invaluable and irreplaceable resources in various forms, ranging from the tangible, such as financial support and volunteerism, to the less easily measured, such as a collaborative spirit and an unrelenting sense of optimism. Intersecting forces of federal and state policy, determined and unwavering support from a diverse set of stakeholders, and widely differing perspectives for solving problems present challenges that can only be solved from transparency, from collaboration, and from communication.

In *Crisis Communication Centers*, Bolanle Olaniran and Juliann Scholl develop very strong arguments for how a diverse body of individuals and organizations can collaborate and coordinate their activities in pursuit of risk and crisis management. These centers are based on the need for a broader, more systemic and interdisciplinary approach for understanding how communities manage risks, crises, and disasters. Properly conceived and operationalized, the centers will serve as the nexuses for the next generation of research and education for promoting community safety and health. In addition to the practical implications discussed by Olaniran and Scholl in Chapter 10, additional opportunities for the crisis communication centers (CCC) could be promoting multidisciplinary programs in risk sciences, translating science into practice, minimizing risks of vulnerable populations, improving evidenced-based decision making under uncertainty, and fostering interdisciplinary risk management communities. According to the authors, "The uniqueness of the CCC rests in the fact that it is a repository of information, evaluation or analysis, and communication training that crisis practitioners in local and surrounding communities can utilize" (Chapter 2). Done well, the CCCs will provide a fundamental infrastructure to facilitate rapid advances in the following vital areas:

- warning systems
- protection of vulnerable populations
- risk information management (Informatics)
- community infrastructure building
- decision making under uncertainty
- interorganizational and interdisciplinary collaboration
- communication of science and technology
- emergency and crisis communication plans

An initiative involving multidisciplinary collaboration focusing on society health and safety is an opportunity to elevate research and professional processes, and to ask new questions about risk and crisis dimensions. These problems require large investigative systems, which need expert staff seldom found in single departments. In addition, these collaborative initiatives should represent more than a collection of involved parties in search of funding opportunities. Instead, they should be constituted as places where new ideas are incubated and nurtured through collaboration stimulated by bridging various aspects of disciplines that create a critical mass of interdisciplinary activity.

The CCCs' efforts will be used to improve the effectiveness and efficiency of risk and crisis mitigation, response, and recovery. They will also serve as a mechanism for advancing interdisciplinary exchange on issues related to natural, technological, and man-made threats and adverse events, especially in relation to vulnerable population groups. As envisioned by Olaniran and Scholl, the CCCs are uniquely positioned to develop evidenced-based, predictive science models for preventing, avoiding, mitigating, and recovering from adverse situations of natural, technological, or man-made disasters. In organizing their presentation, Olaniran and Scholl took special steps to emphasize anticipatory models of crisis response and management (Chapter 1); from there, they identified key characteristics inherent in CCCs, such as information and communication technology (Chapter 4), and issues involving culture (Chapter 7) and ethics (Chapter 8). Traditional media and key stakeholders were also given special attention. The book concludes with a plan of action (Chapter 9) and practical implications (Chapter 10), thus coming full circle to the critical points advanced in the Introduction—protecting communities in a collaborative manner.

The notion of CCCs is fostered by previous and existing attempts to bring community organizations together. Whether they be supported by authority-driven organizations, such as Local Emergency Planning Committees, or instigated by more conceptual notions, such as community infrastructure models (O'Hair, Kelley, & Williams, 2011), these CCCs will play a key role in information exchange and lead to appropriate messaging to the public and the coordination of strategies for prevention, response, and recovery efforts.

A communication perspective assumes the roles and responsibilities for engaging in enlightened conversation as people collaborate and compete in the definition and operationalization of crises. The last two decades observed a proliferation of crisis communication research focusing primarily on serious diseases, natural and man-made disasters, and a multitude of hazards. Simultaneously, developments in communication research has reached levels of proliferation unmatched in the history of the discipline (Heath & O'Hair, 2009). A dramatic emphasis on crisis communication events creates unprecedented opportunities to integrate risk and crisis communication practice with new approaches to preparedness and response, thus resulting in innovative models of community crisis communication as purported by this new book.

References

Heath, R. L., & O'Hair, H. D. (2009). The significance of risk and crisis communication. In R. L. Heath & H. D. O'Hair (Eds.), *Handbook of risk and crisis communication* (pp. 5–30). New York: Routledge.

O'Hair, H. D., Kelley, K., & Williams, K. (2011). Managing community risks through a community-communication infrastructure approach. In H. Canary & R. McPhee (Eds.), *Communication and organizational knowledge: Contemporary issues for theory and practice* (pp. 223–43). New York: Routledge.

ACKNOWLEDGMENTS

Bolanle Olaniran

First, I want to thank God—without whom this book never would have happened—for seeing me through this process. I am grateful to my co-author, Juliann Scholl, who worked tirelessly to make this book a reality. I am also grateful to my family, Adeline and Esther-Hope, who provided the support and allowed me the necessary time to work on this book. I want to thank my parents, Samuel and Alice Olaniran. I am also appreciative of support from numerous colleagues, who have allowed me to bounce ideas about this book project off of them. I know I am unable to mention them all, but I would like to thank David Williams, who planted the seed of crisis communication center in our collaborations on crisis preparedness.

Juliann Scholl

I extend my sincere thanks and appreciation to my co-author, Bolanle Olaniran, for his partnership in turning an idea into a reality. He not only has grown to be one of my most trusted colleagues, but he is also my "go-to" for anything on crisis communication, a "big brother," and a dear friend. I also thank David "Doc" Williams for sharing his crisis expertise with me and for being a positive source of support during my early career development. I also wish to thank my mom and dad, Dianne and LaVern Scholl, for their support and the work ethic they instilled in me. I owe many thanks to Diane Papes, a mentor and dear friend; she has stood by me through the best and worst times and has inspired me to persevere. Finally, I owe a tremendous debt to the friends, colleagues, and mentors who have inspired me professionally during the course of this project—Tom Cunningham, Rebecca Guerin, Patrick Hughes, Lauralynn McKernan, H. Dan O'Hair, Sandra Ragan, K. David Roach, Paul Schulte, Toby Smith, Robert Stewart, Carol Stephenson, Ravi Vadapalli, Donna Van Bogaert, and Shawn Wahl.

INTRODUCTION

Overview of the Crisis Communication Center

Chapter Highlights

- The nature of crisis as a premise for a Crisis Communication Center (CCC)
- Crises at the community level
- Community preparedness for crises
- The need for a CCC

The 9/11 attacks, school shootings at Sandy Hook Elementary School (Newtown, Connecticut), and the Aurora, Colorado, movie theater shooting in 2012, along with the 2015 church shootings in South Carolina, reminded the American public and the rest of the world that we are at risk and that preparation for crises, disasters, and other inevitabilities is not an option but a necessity. Therefore, if preparation is the key to the continued safety of life (and property) as we know it, then steps must be taken to prevent similar attacks and disasters. Toward this end, the 9/11 attacks led to the creation of the Department of Homeland Security (DHS), which has since given us the color-coding risk level systems, or what the U.S. media fondly refer to as the "terror alert level," that is, the dissemination of information regarding the risk of terrorist acts against federal, state, and local authorities and against the American people.

In spite of such a proactive approach to fighting terrorism abroad and increasing security on the home front, there still exists the fear of future attacks. Evidence also suggests that terror cells and the likes of Al-Qaeda, ISIS, etc. are getting increasingly innovative in trying to beat the security systems in place. For example, the shoe bomber and the underwear bomber remind us that the risk is real, and this risk has since informed our way of life. Public concern appears to escalate when the terror alert level is increased—when significant national holidays or events approach or on the anniversary of 9/11, for example. Similarly, specific areas of the country suffer anxieties stemming from the proximity of potential terrorist targets at the local level.

Furthermore, although not acts of terror, natural disasters, such as hurricanes Katrina and Rita, the Japanese tsunami, the 2010 Brazil flooding, and the earthquake in Haiti, remind us that nations, organizations, and local communities must focus their attention on adequate preparation to prevent or to reduce the impact of crises or disasters. As if terrorism and natural disasters are not rampant enough, organizational mishaps, technology failures, and human errors all contribute to crises of unimaginable proportions. For instance, the 2010 Deepwater Horizon oil spill, referred to as the BP blowout, resulted in one of the largest environmental/ecosystem crises of our time, similar to, or greater than, the 1989 Exxon *Valdez* disaster.

When a crisis happens, regardless of the nature or cause, community leaders and members generally want to be informed about the type of crisis event and how to respond appropriately. Attention is drawn to the initial impact of such crisis events, but what often gets overlooked is the anticipation of such events and appropriate communication with the public to ensure that such impacts are minimized. To this end, communities need to have support systems in place to help them respond and communicate appropriately.

Crisis communication plays an important role in maintaining a community's safety and security. Equal in importance to responding properly to crises is anticipating them and preparing for communication about these events. Even though governments devote significant attention to national crises, anticipation and preparation specific to local communities are imperative and can assist media outlets, elected officials, and message designers in reaching their intended target audiences with the utmost success. However, community leaders and members of the media might not possess the communication skills and knowledge needed to prepare a local community for potential crises.

Therefore, there is a need for crisis communication centers (CCC) where community leaders can turn for assistance.

CCC Overview

Crisis communication centers are locally planned and developed agencies that can be staffed according to the size of the community they represent. These centers can be locally governed, but they are still connected to DHS. The purpose of such centers is to regularly monitor the community for vulnerability to terrorist threats, acts, and crisis vulnerabilities, while facilitating communication with the public regarding those concerns. CCCs can be local government agencies, but their success depends greatly on local community and industry support and on the availability of some media, communication, and public relations expertise (Scholl, Olaniran, Williams, & Heuman, 2008; Scholl, Williams, & Olaniran, 2005).

Crisis community centers are needed in local communities throughout the United States and elsewhere around the world (Scholl, Williams, & Olaniran, 2005, Scholl, Olaniran, Williams, & Heuman, 2008). The CCC is grounded in the anticipatory model of crisis management (AMCM) (Olaniran & Williams, 2001), which stipulates that preventing crises and developing crisis messages require preparation, and that crisis communication functions at its best when it is proactive and adequately planned (Benson, 1988; Coombs, 1999; Olaniran & Williams, 2001; Williams & Olaniran, 1994; Williams & Treadaway, 1992). Within an anticipatory framework, crisis managers and practitioners have a strong understanding of their environments and technologies and of the people and procedures involved.

The university setting represents an ideal place for a CCC, which serves not only the community at large but also instructors, research faculty, and students, along with community leaders who can benefit from training in crisis communication. A CCC can function as a knowledge repository with instant credibility among community members and stakeholders who have the need for information gathering and for training about crises and crisis communication. Furthermore, a CCC can, with the aid of information communication technologies (ICTs), allow other researchers and practitioners access to information, and it can promote curriculum development and enhancement through e-learning (Olaniran, 2009). For instance, collaborative learning and training can occur via audio and video conferencing for individuals who are not co-located, allowing them to jointly participate in training and development or distance learning, along with consulting. Training and curricula do not have to be confined to a particular location and thus could be truly global. Finally, the CCC can help provide a unique way to develop both local and nonlocal access to information, either through training or general development about crisis anticipation and preparedness.

CCCs—with their foundations in the anticipatory model of crisis management—ensure that crisis managers optimize their success in responding to crises when they vigilantly anticipate the potential for crises. The process involves a detailed analysis of all of the factors related to specific communities, including (but not limited to) environment, technology risks, human errors, and consequential financial vulnerabilities. Furthermore, the AMCM encourages crisis experts and practitioners to develop and engage in systematic review of all crises concerns, to provide response strategies, and to maintain relationships with stakeholders who can assist with crisis responses. A thorough reconnaissance of each crisis concern is called for in an attempt to develop specific and appropriate crisis responses and management. It is believed that the greater the crisis response, the greater control crisis experts will have in addressing a particular crisis (Olaniran & Williams, 2001, 2004; Scholl, Williams, & Olaniran, 2005).

Staffing CCCs

For CCCs, anticipation is crucial in the signal detection of terrorist threats and acts. CCCs are locally developed and maintained units that function to assess terrorist vulnerabilities and to decide the optimal means of communicating terrorist threats and acts to their specific communities. They are staffed by community members and by municipal employees who have specific knowledge and expertise within the community. The staff can share their expertise about the community in determining what terrorists would most likely target. For example, in a farming community, the CCC would need input from agricultural experts in the area, who could assess how terrorists might attack their industry. Risks such as the destruction to crops in order to attack the economic base of the community, the tampering with livestock to spread illness, or the stealing of crop-dusting planes in order to use them to spread dangerous chemicals would all be possible vulnerabilities that local agricultural experts might address (Scholl, Williams & Olaniran, 2005; Scholl, Olaniran, Williams, & Heuman, 2008).

Choosing appropriate staff for CCCs is crucial. It is important that experts be selected from within the community. For instance, members of the health profession, engineers, and crisis communication scholars are key representatives on CCC staff; also important are stakeholders with whom the CCC staff maintains contact. Given that most potential terrorist threats involve a concern for health and safety, these stakeholders must be aware of their roles in crisis responses and they must know how to manage their communication with the public. Their roles move beyond treating the aftermath of a terrorist attack. They also play a critical role in information gathering, analysis, dissemination, and community awareness (Scholl, Williams, & Olaniran, 2005). The respective experts are consulted for input; for example, the CCC needs to work closely with members of the health profession to create messages that alert the public to health concerns while also helping to alleviate public concerns and prevent escalating panic. Responsibility for communication about public safety concerns from the health profession reaches beyond the hospital spokesperson and to individuals actually working in a hospital with victims of an attack.

Once appointed and assembled for the onerous responsibility, CCC staff begin immediate assessment of crisis vulnerabilities and terrorist risks in their community. Guidelines from DHS provide a starting point for this process, as the team identifies targets, risks, and vulnerabilities. However, team members' local expertise and connections in the community provide the most valuable insights. The CCC develops and maintains a list of areas of concern in the community and maintains relationships with key individuals who can alert the CCC to changes or to activities that might signal a threat. For example, early insight from the community hospital of otherwise unnoticeable, but unexplained, injuries due to minor chemical burns or illnesses from an unknown vapor might signal the onset of a terrorist act that would otherwise go undetected (Scholl, Williams, & Olaniran, 2008). From a different standpoint, the Lubbock, Texas, hepatitis A threat offers a point of reference. An employee at the local Cheddar's Casual Café contracted the virus and potentially exposed anyone who ate at the restaurant. Consequently, the city's health department sent letters to those at risk and urged them to get vaccinated. When it was all over, 2,700 individuals were vaccinated, and the vaccines cost the city about $250,000, but individuals were asked to pay $15 each to recoup some of the cost. The good news is that because of collaboration between health professionals and the city, no new case was reported beyond the initial one (Hepatitis Attorney, 2012), and a major epidemic was averted.

Public relations and crisis management personnel are advised to have regular crisis response training. CCC staff ideally also have regular reviews of vulnerabilities and periodic communication with their contacts within the community. The development of a community-assessment instrument has been suggested elsewhere as a means for facilitating this process (Scholl, Williams, & Olaniran, 2005; Scholl, Olaniran, Williams, & Heuman, 2008). The instrument might include a generic inventory of known areas of vulnerabilities and a list of local sites of interest. In other words, the aim of the CCC is to identify and to research potential local terrorist targets, crises, and other disasters, while developing the best means of information dissemination for the particular community. Such activities are conducted with the belief that the anticipation of crises and/or terrorist attacks and the preparations for them at the local level can supplement national efforts to keep people safe. From this standpoint, community members worry less about their safety, because they can monitor only the national alert systems of interest to them.

Identifying targets and crisis preparation for CCCs is similar to standard procedures for crisis management teams. However, specific needs of CCCs require early attention to the completion of a community assessment. Several types of potential crisis sources and terrorism targets exist in the United States. Perhaps at the highest risk for attack are the nation's 361 commercial ports, through which $750 billion in goods flow each year (Bender, 2003; Scholl, Williams, & Olaniran, 2005; Seikman, 2003). Bender (2003) reports that the United States is more likely to be attacked with a weapon of mass destruction (WMD) smuggled through a port than with a missile. Other devices that could easily be smuggled include radioactive or biologically malignant materials. With security heavily focused on air travel, the lack of funding to bolster seaport security makes our ports especially vulnerable. Besides the danger of smuggled weapons, an attack that temporarily shuts down port operations could restrict access to such items as spare parts for major industries and heating oil needed in winter (Bender, 2003).

Locations known for marquee events or tourist attractions are also at risk. For example, tempting potential terrorist targets include New Orleans, with its Mardi Gras festival; New York, with its New Year's Eve celebration at Times Square; and other such regular celebrations. Tourist spots, such as Disneyland and casinos in Las Vegas, have been targeted by "sleeper cells" whose members collected intelligence on these areas (Runk, 2003) as targets of future terrorist attacks (Scholl, Williams, & Olaniran, 2005; Scholl, Olaniran, Williams, & Heuman, 2008). Famous landmarks, such as the Golden Gate Bridge in San Francisco, might also be targets (Scholl, Williams, & Olaniran, 2005; Zamora, 2003).

It is noteworthy that while some targets might seem obvious to the public, any location housing government facilities, energy and transportation infrastructures, prominent financial and commercial districts, and agricultural interests are also potential targets (Azar, 2002). Widely dispersed chemically based threats, such as bioterrorism and environmental terrorism, means that highly populated states, such as California, New York, and Texas, are not the only terrorist targets. Seemingly unlikely states, such as Wisconsin and Rhode Island (Barbarisi, 2003; Williams, 2003), are especially vulnerable because of the nuclear power plants and factories in their state (Scholl, Williams, & Olaniran, 2005; Zoretich, 2003). In actuality, no community should consider itself immune or too underpopulated to be at risk of a terrorist attack and crisis. Predicated on principles of the AMCM, CCCs must consider, in particular, the ease with which perpetrators can target the general public by creating a gas attack, by poisoning a food supply, by disrupting an energy grid, by killing large numbers of livestock, or by contaminating a natural resource (Azar, 2002; Chalecki, 2002).

The goal of identifying crisis potentials is to develop a list that can be prioritized and categorized according to conceivable risks. For instance, it has been suggested that a crisis review should include historical references to past events in an attempt to study the pattern and lessons learned toward lowering subsequent potential crises (Hubbard, 2009; Lerbinger, 2012). At the same time, Lerbinger (2012) recommends that after the list of potential targets has been identified, each crisis can be categorized based on the level of risk (e.g., minor, serious, and major). He explained a minor-level crisis as one causing only few injuries and damage to third parties and to the environment. He also emphasizes that in most cases, minor crises may not be of interest to the media. The serious crisis category, on the other hand, is one that might occur within and outside an organization or community, but it is usually controlled and contained. It also will involve authorities in and outside the community and might also moderately interest the media. The major crisis category includes incidents where control or containment has not yet been established, and they usually result in injuries and fatalities, involve authorities, and have public significance and media attention (Lerbinger, 2012). These three categories can be color-coded similarly to those in place at the national level by DHS or in any state department crisis-readiness alert system.

The Role of the Media in a Crisis

The media are the disseminators of important crisis-related information, and they play a significant role in the success or failure of crisis management. When using the media to influence public opinion, crisis managers must develop meaningful cooperation with the media and with their respective reporters. However, as the dawn of the twenty-first century ushered in new media outlets, social media are redefining what counts as media and how individuals prefer to receive information. For instance, it is not uncommon to find information about riots and political uprisings on Twitter and on Facebook as they are happening, even before the traditional mass media (e.g., radio and television) are able to cover them (Lerbinger, 2012). Although some criticism has been levied against the mass media on the basis of their increased coverage of political violence and portrayal of crises and disasters, it needs to be said that the media provide a valuable function to the public in times of crises (Scholl, Williams, & Olaniran, 2005). Similarly, publication of the National Academy of Sciences report, *Disasters and the Mass Media*, brought to light many relevant roles the media can play when disasters strike, especially in educating the public by offering warnings and directions for actions and by disseminating information on the appropriate agencies to contact for assistance (Tierney, Lindell, & Perry, 2001). Apropos are the events following the 9/11 terrorist attacks, which appeared to bring the public and the news media closer together as news reporting delivered comfort to individuals, provided urgent information, helped the public take appropriate courses of action, and directed people on what to do in public spaces (Nacos, 2003; Scholl, Williams, & Olaniran, 2005; Scholl, Olaniran, Williams, & Heuman, 2008).

According to Nacos (2003), the constant flow of information provided by print, radio, and television made the public feel more involved with the unfolding events of 9/11. Also, the news media were seen as a valuable resource in assisting crisis managers with communicating important information (e.g., what to do and what not to do). Finally, the dissemination of news provided individuals with public spaces where they could converse with experts and with each other about the terrorist acts, their aftermath, and their implications (Scholl, Williams, & Olaniran, 2005). Another study suggests that the media played five functional roles as perceived by the public: (a) providing information, (b) building solidarity, (c) providing comfort, (d) explaining the significance of an event, and (e) reducing tension. These five roles are considered by respondents to be very important in the aftermath of a crisis, particularly a terrorist attack (Perse et al., 2002).

For many people and in most situations, the often centralized mass media are the most salient sources of information on hazards and disasters, and they in large measure helped people to learn what they know about disasters (Tierney, Lindell, & Perry, 2001). Recently, however, individuals are increasingly engaging new social media, such as Twitter and Facebook, as a way to stay informed and also to develop a sense of community where one can reach out to discuss crises events while letting out feelings and emotions, and personalizing or evaluating certain events and experiences (Greenberg, Hofschire, & Lachlan, 2002). Because of the wide reach and immediacy of television, we now expect our policymakers and crisis managers to be on top of emerging events. Selection of an effective medium for communicating with the public is another vital role of the CCC. Qualified staff members should survey the community to answer some basic questions about their information-gathering preferences. Foremost among concerns over information dissemination is who should present the information, what types of media community members prefer or use, and whether there is an individual (other than the CCC) spokesperson to whom the community would be more attentive. In other words, is there a credible person whom citizens would prefer to deliver news about threats in their community? The CCC must also determine what level of potential threat warrants public communication. Answers to these questions can vary greatly depending on communities, and this highlights the need for locally developed and operated CCCs (Craig, Olaniran, Scholl, & Williams, 2006; Olaniran, 2007; Scholl, Olaniran, Williams, & Heuman, 2008).

For CCC teams, members of the local government should be included as spokespersons. An ideal guideline was established by Mayor Rudy Giuliani in New York City in the wake of 9/11. Ideally, the local mayor or city manager should be able to serve in this capacity for most communities. Other individuals would be recognizable members of the city council, members of industry, or key leaders in the health or public safety sector. In some communities, a clergy member may be a good spokesperson. Criteria for CCC spokespersons, as with most crisis management spokespersons, include the ability to work with the media. Aside from public speaking skills, these individuals will need to respond to questions with accuracy, tact, and composure. The latter might prove to be difficult, as spokespersons could be called upon in times of great stress. Therefore, Coombs's (1999) suggestion that spokespersons receive media training is wise, and that training could be offered by CCC staff.

CCCs, organizations, and agencies with the responsibility to protect and serve the community must understand the media's impact on the public's perception of a crisis and circumvent adverse outcomes from news coverage, but they must learn to disseminate messages that can assist the community during a crisis (Olaniran, 2007; Olaniran & Williams, 2004; Scholl, Williams, & Olaniran, 2005). They may even have to engage in some level of data-mining to determine how the public and different demographics select media and acquire information (see Craig, Olaniran, Scholl, & Williams, 2006; Olaniran, 2007).

Research examining the relationship between the media and the public suggests that the public consists of multiple audiences, and that any message communicated during a crisis must have a clear goal understandable to each audience (Albrecht, 1996; Olaniran, 2007; Scholl, Williams, & Olaniran, 2005). Otherwise, different audiences might perceive different messages and interpretations from the crisis leaders, who must minimize the potential of disclosing damaging information or of sending the wrong message to the wrong audience. The potential damage is greatest with television coverage, as politicians have developed a much closer relationship with reporters from television than from other media (Crelinsten, 1994; Seeger, Vennette, Ulmer, & Sellnow, 2002; Stempel & Hargrove, 2002).

Crisis managers and decisionmakers must anticipate the results of unexpected or unintended information feedback loops and cultivate an awareness of where messages might go versus where they are intended to go (Crelinsten, 1994). In light of the potential impact of a local crisis on the national or international community, managers need to adapt their behaviors and decision making to the particular demands of the electronic media.

Technological elements of public communication will be another component of the CCCs' work. Use of the Internet to host a website, to send e-mail, and to host bulletin boards will be central to the work of information dissemination. The primary purpose of such websites would be to facilitate communication with the public by both receiving and sending information. For example, local CCCs could sponsor websites that would allow citizens to respond to inquiries regarding from whom they would like to receive crisis information and how they would like to receive those messages. This could prove to be useful in the early stages of the development of each CCC as it works to determine how to present information to the public and who should present it (Scholl, Olaniran, Williams, & Heuman, 2008). In taking advantage of the utility of social media, CCC communications staff could establish and maintain Twitter accounts in order to follow feeds and updates, which could shed more light on community members' expressed preferences.

In times of crisis or uncertainty, electronic communication could be used to alert the public about what is happening; what the public should do; and what the perceived risks might be. Access to this information would be an issue of concern for each local team. Electronic communication could also be used for internal purposes by members of the team.

Coordinating the communicative elements to serve the diversity of interests and abilities represented in each local community would take extensive work on the part of CCC teams. Tools such as video conferencing to connect with expert individuals located in different geographic locations and access to authoritative figures in government agencies are worth exploring. Particular attention must be given to the role of mediated public relations as it interfaces with health professionals; to members of society who have less influence, input, and access to public information; and to those who understand the complexities of technological issues in computer-mediated communication. CCCs should meet both physical and equipment requirements (Lerbinger, 2012). The size should be larger than what is required to accommodate expected numbers of media representatives. It is recommended that the space consists of a separate room for interviews, which should contain tables and appropriate furniture. In addition to tables and furniture, the center must be housed with adequate communication technologies, computers, copying equipment, fax machines, and miscellaneous supplies (e.g., paper, pencils and pads, staplers, reference books, telephones, communication radios—especially when mobile phone lines are disrupted or not available—telephone directories, video and audio equipment, coat racks, and chalk boards). Press kits are encouraged and should be stored digitally, and there must be adequate servers to prevent website crashes as stakeholders overwhelm the centers in search of information about crises. The time of a crisis is not the time for a website to crash (Auffermann, 2009). Other logistics that are not required, but might be of great benefit, include a shower room, kitchenette, an adequate supply of water, and backup generators to assist workers who might need to work round the clock or to pull late-shift duties depending on the crisis and the nature of the disaster. For example, tornadoes usually disrupt power lines, which create blackouts and also disrupt the food supply chain. A woman recounted on a television show (*Undercover Boss*, aired March 2, 2012) that in the hours immediately following Hurricane Katrina there was no water supply for 48 hours, and that she had to resort to drinking water from vegetable cans.

Developing a contingency plan or advance arrangement is critical to activities of the CCC. It is imperative that advance arrangements be made to secure permits when required from government agencies. For example, the Exxon *Valdez* disaster made this issue crystal clear (Williams & Olaniran, 1994). Lerbinger (2012) also stresses that if an advanced arrangement is not possible, clearance procedures should be outlined with contact names, addresses, and phone numbers of parties involved.

The entire CCC staff must be made aware of, and understand their roles in, crisis prevention, management, and control. Those likely to receive inquiries about a crisis must have media training and learn to refer inquiries from the media to the CCC. Lerbinger (2012) also suggests that if crisis preparedness involves emergency teams, the readiness of the teams must be incorporated into rehearsals and should be carefully monitored. Regular crisis games, simulations, and rehearsals are called for to get the crisis management team into a mode where crisis plans can be implemented as a routine activity. The development of routines will prevent human errors that can further escalate crises (Coombs & Chandler, 1996; Lerbinger, 2012; Olaniran & Williams, 2001, 2004; Scholl, Williams, & Olaniran, 2005). Of course, crisis rehearsals also serve as a way to spot problems in a given crisis plan and to put in place efforts to correct the problem—in particular, when a key staff person is missing.

Conclusion

The CCC provides a place where community resources can be coalesced and analyzed for appropriate implementation in the case of a crisis. It also presents a credible source from which experts and staff can identify targets that are vulnerable to crises and disasters and put in place plans for addressing them. More importantly, the CCC is a place where team members can function collectively to make their communities safe for citizens and for residents and a place to train crisis managers of tomorrow.

References

Albrecht, S. (1996). *Crisis management for corporate self-defense*. New York: American Management Association.

Auffermann, K. (2009, Summer). It still chokes me up—Responding to a deadly campus shooting at Northern Illinois University. *The Strategist*, *15*, 14.

Azar, L. (2002, August 8). 27 sites could be terrorist targets, subcommittee finds. *Wyoming Tribune–Eagle*, p. A4.

Barbarisi, D. (2003, May 23). Terrorism in RI: You never know. *Providence Journal Bulletin*, p. D4.

Bender, B. (2003, June 21). Terror risk seen highest at US ports: Tankers in Boston may be vulnerable. *Boston Globe*, p. Al.

Benson, J. A. (1988). Crisis revisited: An analysis of strategies used by Tylenol in the second tampering episode. *Central States Speech Journal*, *39*, 49–66.

Chalecki, E. L. (2002, February 21). A new vigilance: Identifying and reducing the risks of environmental terrorism. *Global Environmental Policy*, *21*, 46–64.

Coombs, W. T. (1999). *Ongoing crisis communication: Planning, managing, and responding*. Thousand Oaks, CA: Sage.

Coombs, W. T., & Chandler, R. C. (1996). Crisis teams: Revisiting their selection and training. In L. Barton (Ed.), *New avenues in risk and crisis management*, *Vol*. 5 (pp. 7–15). Las Vegas: University of Nevada, Las Vegas Small Business Development Center.

Craig, M., Olaniran, B. A., Scholl, J. C., & Williams, D. E. (2006). Crisis communication in public arenas. *Public Relations Review*, *32*(2), 171–73.

Crelinsten, R. D. (1994). The impact of television on terrorism and crisis situations: Implications for public policy. *Journal of Contingencies Crisis Management*, *2*(2), 61–72.

Greenberg, B. S., Hofschire, L., & Lachlan, K. (2002). Diffusion, media use and interpersonal communication behaviors. In B. Greenberg (Ed.), *Communication and terrorism: Public and media responses to 9/11* (pp. 3–16). Cresskill, NJ: Hampton.

Hepatitis Attorney (2012, February 19). Lubbock's Cheddar's Casual Café costs public health $250,000 due to hepatitis A. Retrieved February 22, 2012, from http://www.hepatitisblog.com/hepatitis-a-watch/.

Hubbard, D. W. (2009). *The failure of risk management: Why it's broken and how to fix it*. Hoboken, NJ: John Wiley.

Lerbinger, O. (2012). *The crisis manager* (2nd ed.). New York: Routledge.

Nacos, B. L. (2003). Terrorism as breaking news: Attack on America. *Political Science Quarterly*, *118*(1), 23.

Olaniran, B. A. (2007). The role of perception in crisis management: A tale of two hurricanes. *Multicultural Education*, *15*(2), 13–16.

Olaniran, B. A. (2009). Organizational communication: Assessment of videoconferencing as a medium for meetings in the workplace. *International Journal of Human Technology Interaction*, 5(2), 63–84.

Olaniran, B. A., & Williams, D. E. (2001). Anticipatory model of crisis management: A vigilant response to technological crises. In R. Heath (Ed.), *Handbook of Public Relations* (pp. 581–94). Thousand Oaks, CA: Sage.

Olaniran, B. A., & Williams, D. E. (2004). Burkian counternature and the vigilant response: An anticipatory model of crisis management and technology. In D. Millar & R. Heath (Eds.), *Responding to crisis: A rhetorical approach to crisis communication* (pp. 75–94). Mahwah, NJ: Lawrence Erlbaum.

Perse, E., Signorielli, N., Courtright, J., Samter, W., Caplan, S., & Lambe, J. (2002). Public perceptions of media functions at the beginning of the war on terrorism. In B. Greenberg (Ed.), *Communication and terrorism: Public and media responses to 9/11* (pp. 39–52). Cresskill, NJ: Hampton.

Runk, D. (2003, June 4). Two found guilty of scouting terrorism targets: "Sleeper cell" had eye on Disneyland. *The Record*, p. A12.

Scholl, J. C., Olaniran, B. A., Williams, D. E., & Heuman, A. (2008). Data collection for crisis communication centers: Community preferences for terror threat and act information. In A. Merkidze (Ed.), *Terrorism issues: Threat assessment, consequences, and prevention* (pp. 153–74). New York: Nova Science Publishers.

Scholl, J. C., Williams, D. C., & Olaniran, B. A. (2005). Preparing for terrorism: A rationale for crisis communication center. In H. O'Hair, R. Heath., & G. Ledlow (Eds.), *Community preparedness and response to terrorism* (pp. 243–68). Westport, CT: Praeger.

Seeger, M. W., Vennette, S., Ulmer, R. R., & Sellnow, T. L. (2002). Media use, information seeking, and reported needs in post crisis contexts. In B. Greenberg (Ed.), *Communication and terrorism: Public and media responses to 9/11* (pp. 53–63). Cresskill, NJ: Hampton.

Seikman, P. (2003). Protecting America's ports. *Fortune*, *148*(10): TI98C.

Stempel III, G. H., & Hargrove, T. (2002). Media sources of information and attitudes about terrorism. In B. S. Greenberg (Ed.), *Communication and terrorism: Public and media responses to 9/11* (pp. 17–26). Cresskill, NJ: Hampton.

Tierney, K. J., Lindell, M. K., & Perry, R. W. (2001). *Facing the unexpected: Disaster preparedness and response in the United States*. Washington, DC: Joseph Henry Press.

Undercover Boss (2010, March 2). CBS TV Show.

Williams, D. E., & Olaniran, B. A. (1994). Exxon's decision-making flaws: The hypervigilant response to the *Valdez* grounding. *Public Relations Review*, *20*, 5–18.

Williams, D. E., & Treadaway, G. (1992). Exxon and the *Valdez* accident: A failure in crisis communication. *Communication Studies*, *43*, 56–64.

Williams, S. (2003, February 8). State cut of funds bigger than most: $8.7 million received to boost anti-terror arsenal. *Milwaukee Journal Sentinel*, p. Bl.

Zamora, J. H. (2003, February 23). California lists top terror targets. *San Francisco Chronicle*, p. A1.

Zoretich, F. (2003, July 12).Terror threat great, but U.S. more ready, Freeh says. *Albuquerque Tribute*, p. A5.

PART I
THEORETICAL UNDERPINNINGS

· 1 ·

THE ANTICIPATORY MODEL OF CRISIS MANAGEMENT AND THE CRISIS COMMUNICATION CENTER (CCC)

Chapter Highlights

- Making the case for crisis planning and preparedness
- Why the crisis communication center?
- How CCCs differ from other crisis response centers
- The AMCM theoretical perspective and the CCC

If there is one certainty, it is that crises are inevitable. Crises almost always occur at the most inopportune time. The times in which we live suggest the need to be prepared for crises of all forms, including disasters and terrorist threats. It is no longer a question of whether crises will happen but when. Four months into 2015, there were three outbreaks of the norovirus on cruise lines, and nine in the previous year. Prior to that, there was the spread of Ebola to the United States and to several European nations. Then there was the BP Deepwater Horizon drilling rig explosion that saw several million barrels of oil spilled after a well was drilled in the Macondo Prospect in the Gulf of Mexico. It took more than three months to stop the oil spill.

These examples point to the necessity of crisis planning and crisis preparedness. Among the lessons learned from all of these crises is that planning is paramount and that crisis managers should consider wisely and proactively the likelihood of various issues or risks that can manifest in ways that require prevention, mitigation, and response. In other words, issues and their management are essential to effective crisis planning and to preparedness. Therefore, it is the responsibility of key decision makers and leaders in organizations, communities, and governments to be prepared for the inevitable, because the cost of unpreparedness is significantly greater.

Crisis discussions, however, often feature challenges faced by private-sector organizations. Indeed, there are many iconic cases in businesses and in government and in nonprofit organizations where executives fail their organizations and their stakeholders. Heath and Sultan (2012) discuss how the actions of U.S. presidents mar their reputations and put policy positions at risk (or cause their failure). Members of Congress become involved in ethics scandals. Nonprofit organizations blunder. So, organizations of all types, along with communities across the globe, need pre-crisis management and communication policies and their implementation. This point is further illustrated by Lukasewski (2013), who argues that all crises create victims, and thus communication is required in the early stages of a crisis to address victims' needs.

It is also pressing that crises on the part of one person or organization have a way of resulting from, and creating crises for, others. The general consensus on crises is that they involve *high consequence*, have *low probability*, consist of *high risk and uncertainty*, operate under *time pressure*, interfere with *normal business*, and *potentially* affect *organizational reputations* (Gregory, 2005; Pearson & Clair, 1998). Notwithstanding, Jaques (2006) suggests that the challenge is not so much in defining issues and crises, but in the application of those definitions to management theory and to practice. For instance, a common misconception about crises and, in particular, crisis management is the emphasis on reactions or tactics used once a crisis has occurred, particularly image repair following a crisis. However, crises involve issues management, which is at the center of planning and crisis preparedness. To put this into perspective, one of the very first iconic books on crisis management is by Fink (1986), who penned the expression "Crisis management: planning for the inevitable," hence pointing to the significance of issues in crisis management. Although the role of issues management has generated different interpretations over the years, some scholars have taken it on in their works (e.g., Heath, 2002; Jaques, 2006; Olaniran & Williams, 2001). Another layer of complexity exists for the practical application of issues in crisis management because of the different approaches in academic settings. To this end, we call for crisis communication centers.

What Is a Crisis Communication Center?

A CCC is a community-based center that consists of necessary resources required for the preparation and provision of assistance in the management of all forms of crises and disasters. Crisis centers exist in abundance in North America, Europe, and in other parts of the world. Most tend to have a specific focus. For example, the International Association for Suicide Prevention (IASP) provides links to a network of suicide crisis centers all over the world (IASP, 2015). Other crisis centers provide assistance and information on sexual assault, such as Rape Crisis Network Europe (n.d.). There are also networked crisis response organizations, such as the Emergency Response Coordination Centre (ERCC), which operates within the European Commission's Humanitarian Aid and Civil Protection Department (European Commission, 2015). The purpose of the ERCC is to utilize resources from participating countries to coordinate rapid responses to disasters both in and outside of Europe.

Similar to a network such as the ERCC, a CCC serves as a focal point for gathering information and for monitoring threats and disasters, and it equips citizens and experts to anticipate and deal with crises. The main difference, however, is that the coordination does not occur across a country or region; the coordination hub is the individual CCC within the community. The primary role of the CCC is to function as a knowledge repository for crises and as a communication networking hub for the community. A CCC in such a capacity will possess instant credibility with community members and with stakeholders who have the need for information gathering, analysis, and training about crisis and crisis communication. Furthermore, the center, with the aid of information communication technologies, can allow anyone (e.g., researchers and practitioners) to access information, enhance curriculum development, and implement e-learning (Olaniran, 2007). For instance, collaborative learning and training can occur via audio and video conferencing in a manner where individuals who are not co-located can jointly participate in training and development, or in distance learning with consulting. With this in mind, training and curricula do not have to be confined to a particular location and thus could be truly global. Furthermore, the CCC can help provide a unique way to develop both local and nonlocal access to information, either through training or general development about crisis anticipation and preparedness.

CCCs—with their foundations in the anticipatory model of crisis management—hold that crisis managers optimize their success in responding to a crisis when they vigilantly anticipate the potential for a crisis. The process involves a detailed analysis of all of the factors related to specific communities, including (but not limited to) environment, technology risks, human errors, and consequential loss of life and financial security. Furthermore, the anticipatory model of crisis management (AMCM) encourages crisis experts and practitioners to develop and to engage in a systematic review of all crisis concerns, to provide response strategies, and to maintain relationships with stakeholders who can assist with crisis responses. A thorough reconnaissance of each crisis concern is called for in an attempt to develop specific and appropriate crisis response and management. It is believed that the greater the crisis preparation plan, commitment, and response, the greater control that people managing a crisis will have in addressing a particular crisis (Lukasewski, 2013; Olaniran & Williams, 2001, 2004; Scholl, Williams, & Olaniran, 2005).

The uniqueness of the CCC rests in the fact that it is a repository of information, evaluation or analysis, and communication training that crisis practitioners in local and surrounding communities can utilize. The CCC also serves a unique function that is different from other organizational-level crises in that it is not tailored to address a particular crisis, but it rather provides resources and communication training for any type of crisis or disaster. This is important given that most scholars recognize and call for a greater role of communication in crisis management (e.g., Coombs, 1999; Heath & Sultan, 2012, Lukasewski, 2013; Olaniran, Scholl, Williams, & Boyer, 2012). However, there is no consensus on how to bring this about. This book, through its proposition of the CCC, attempts to fill that void. The CCC provides an advisory role to communities, organizations, and individuals alike, who are interested in the idea of effective crisis or disaster management, want access to critical and essential information, and are in need of the expertise provided by the center. Also, CCCs can allow stakeholders and community members to take active roles in the crisis management process. For example, crises often provide physical cues or early warnings that can prompt decision-makers to take actions (Jaques, 2012; Lindell & Perry, 2004, Olaniran & Williams, 2001; Olaniran, Williams, & Coombs, 2012). Thus, such cues, environmental or otherwise, when spotted by community members, can be reported to the CCC, and experts at the CCC can then mobilize actions or contact the appropriate agencies to address the issue. By doing this, individuals and community members are empowered to participate, and, more importantly, they do not have to worry about where and to whom to report certain issues.

The Anticipatory Model

To this end, the AMCM (Olaniran & Williams, 2001, 2004) is a pragmatic tool for those charged with the task of crisis readiness or management, terrorism, and disaster prevention and management. First, however, we need to understand the model. Thus, the following subsection presents an analysis of the AMCM.

The anticipation factor and technology: The AMCM focuses on planning and prevention as primary concerns in the process of crisis and disaster management. The anticipatory model draws from the "enactment perspective" (i.e., Weick, 1988), which focuses on the prevention of possible crises, including disasters, in an attempt to reduce their magnitude when they happen. In essence, the anticipatory model focuses on the ability to predict the unpredictable and to attempt to put in place programs that safeguard against their occurrence.

The AMCM suggests the need for crisis teams and practitioners to pay attention to antecedent factors and, in particular, to technologies that are often viewed as fail-safe (Olaniran & Williams, 2001; Olaniran & Scholl, 2014). When applying the model to crisis communication centers, it is necessary to understand the concept of first- and second-order effects of technology. The first-order effect of technology addresses the notion of technology being used for what it is originally designed (e.g., e-mail being used to send information to a large number of people in a short amount of time). On the other hand, the second-order effect of technology addresses the adaptive and often negative use of technology (e.g., e-mail being used to send a virus over computer networks). For the most part, an act of terror reflects the latter, but a crisis, by its nature, can and might result from both first- and second-order levels. A case in point is 9/11, when terrorists used jet airplanes as high-explosive bombs to cause destruction at the World Trade Center and at the Pentagon. Recently, the Germanwings airline crash disaster by a suicidal pilot is a telltale signal of second-order effect technologies. In essence, the very technology designed to foster better living standards can be used to disrupt lives, environments, and communities. Lerbinger (2012) alluded to this problem when he discussed the intricately complex nature of technologies and processes that embody risks that are increasingly difficult to estimate or to predict.

Consequently, more attention should be devoted to understanding technologies and to anticipating other factors that hold crisis potential or that have negative consequences in an attempt to prevent crisis occurrence and escalation. Furthermore, technologies by nature are fallible because they are designed by humans and are subject to both human errors and deliberate attempts to cause disasters. Evidence indicates that a majority of industrial disasters are linked to a combination of human and technological errors (Jacques, 2007; Lukasewski, 2013; Olaniran & Scholl, 2014; Olaniran & Williams, 2001; Shrivastava, 1987; Shrivastava & Mitroff, 1987; Shrivastava, Mitroff, Miller, & Miglani, 1988; Weick, 1988).

Research over the past several years has provided practitioners with insights into how to identify areas more susceptible to crises and with specific information on how best to manage different crisis types. For example, an organization might be equally susceptible to crises involving natural disasters and organizational misdeeds. A company in California designed to handle hazardous waste cleanups could determine that it needs to prepare for an earthquake that could damage waste containers, thus creating a hazardous waste spill. Additionally, the organization might see worker or management shortcuts that cause environmental threats as a possible crisis risk. Each of these crisis types (natural disaster and organizational misdeed) is plausible and must be anticipated both in terms of management actions and in crisis communication strategies. However, the same preparations and communication strategies would not function optimally for both crisis types. Anticipatory orientation assists public relations professionals and all others involved in crisis and emergency response with the recognition that crisis preparation cannot be performed with one management action or with one response strategy. Rather, crisis management is an ongoing process of monitoring, assessing, and reassessing crisis risks. The practitioner, now more than ever, is asked to look into community, organizational, industrial, and cultural history in order to best prepare for crises. This negates a one-size-fits-all strategy that assumes a single crisis management strategy will work for any community. On the contrary, a community-specific approach—one that is upheld by a CCC—considers what organizational, environmental, and cultural factors unique to a community would most likely fuel crises.

Theory development and model creation is needed to develop the anticipatory orientation with guidance for practitioners and for researchers. Public relations, communication, rhetoric, business management, public health, medicine, emergency response, and a host of other disciplines come into play as scholars and practitioners pull their resources in the development of CCCs that seek to organize and to expand this anticipatory orientation. One of the prominent initial works in this vein was the creation of the AMCM (Olaniran & Williams, 2001). The next subsection offers a brief explanation of the critical components of the anticipatory model.

The AMCM was originally developed to provide assistance to public relations practitioners and researchers who must deal specifically with the threat of technological crises (see Figure 1). The model necessitates an initial understanding and acceptance that technological crises are ever-present risks. Organizational paranoia about technological crises is justified, because previous research has demonstrated and accounted for a significant percentage of reported crises (Mitroff, Pauchant, & Shrivastava, 1989; Olaniran & Williams, 2001). Adopters of technological innovation often fall prey to the perception that materials and design are fail-safe. This lulling of an organization into a sense of ill-founded complacency harkens back to Burke's (1961) warnings of a cultural movement toward Helhaven. Burke warned that as we continue to develop and to depend on technology, we perceive ourselves moving closer to a heavenly state. However, the opposite might actually be more accurate. Incorporating the anticipatory model requires us to accept that greater dependence on technology leads to greater risk. Technology is never foolproof. Even if it was, the human component of technology use is certainly not infallible.

Understanding: Understanding in the AMCM suggests having a thorough knowledge (i.e., reconnaissance) of conditions, situations, or events that could signal the potential for danger. A community-based CCC is ideal for perpetuating understanding because it would be tuned into the conditions and situations specific to the community that could perpetuate a crisis. Understanding cannot be complete without two relatively similar concepts: enactment and expectation. Enactment is defined as a process in which a specific form of action is brought about (see Smircich & Stubbart, 1985). An example of an enactment might be the decision of a company to hire security guards to be posted at entrance gates in response to increased crime in the area. Weick (1988) extends enactment to include consequences imposed by a given action. Olaniran and Williams (2001), however, contend that *anticipation* of a crisis in itself is action because it constrains the choice an organization makes based on given information. Proponents argue that decision makers must look ahead to anticipate opportunities, threats, and weaknesses in their environments and then to take appropriate actions. Furthermore, both action and inaction result in different (i.e., crises) outcomes.

The anticipatory model shares with other theories and models the position that, although human errors cannot be entirely eliminated, the responsibility to engage prevention efforts requiring anticipation of those errors should still exist. Sethi (1987) argues that one should anticipate nearly all of the unthinkables before they become realities (see also Lindell & Perry, 2004). To this end, CCCs are predicated on the AMCM principle that effective crisis preparation and/or management is prevention oriented because it is hard to cope with a crisis for which one has not adequately prepared.

By *understanding*, we mean a thorough working knowledge of the surroundings and factors that can introduce complexities. In particular, the less attention devoted to understanding the nature of a crisis, the more likely the crisis will escalate. In other words, the very action that enables people to develop some understanding of potential crises (i.e., technologies or human errors) can also contribute to or escalate crises (Jaques, 2007; Olaniran & Williams, 2001; Weick, 1988). Consequently, an understanding of the environment in which CCCs serve is essential. A thorough understanding of the environment and the factors that can feed crisis events can transfer into the application of technology. Unfortunately, technologies used to safeguard against disasters are fallible in their own right, simply because they are designed and operated by humans. As a result, fallibility is embedded in these designs. This reasoning finds some justification in that the majority of industrial disasters are linked to technological failures aggravated by human errors (see also Jacques, 2007; Lukasewski, 2013; Sethi, 1987; Shrivastava, 1987; Shrivastava & Mitroff, 1987; Shrivastava, Mitroff, Miller, & Miglani, 1988; Udwadia & Mitroff, 1991; Weick, 1988; Weick & Sutcliffe, 2007). One does not have to look far to notice that a human act is always one of the principal elements that sets crises into motion (Weick, 1988; Weick & Sutcliffe, 2007). The crisis events associated with Three Mile Island, Chernobyl, the Fukushima nuclear plant, and Germanwings airliner are a few of the examples of how establishing CCCs could prevent or minimize the impact of crises. Therefore, a good crisis management model must take into account human errors while anticipating natural occurrences or nonhuman errors that could heighten crisis situations.

Setting understanding into motion with the anticipatory model: Two relatively similar concepts—enactment and expectation—set the tone for *understanding* in the AMCM. Weick (1988) extends the notion of enactment by addressing not only actions but also the consequences imposed by actions. He states, "The term 'enactment' is used to preserve the central point that when people act, they bring events and structures into existence and set them in motion" (p. 306). Enactment constitutes a retrospective sense-making process such that the notion of *anticipation*—or lack of crisis in and of itself—is an action, given that it determines the choice that people make based on available information. Communities and organizations alike must be adept in their decision-making processes; they must anticipate opportunities, threats, and weaknesses in their environment and take appropriate measures to safeguard their interests. Furthermore, it should be noted that individuals' or decision makers' actions (or lack of actions) result in different crisis outcomes. When people act one way versus another, they are faced with different types of problems, opportunities, and constraints (Jaques, 2007; Kilduff, 1996; Lukasewski, 2013; Weick & Sutcliffe, 2007).

Expectation: *Expectation* represents assumptions people make about certain events or objects. Assumptions made about a given situation, environment, and technology, or a combination of all of these, can influence whether or not tragedies occur and determine the magnitude of their occurrences. In essence, assumptions by decision makers about technologies represent critical elements and starting points for preventing terrorism and crises in general. Olaniran and Williams (2001) note that assumptions, at times, bring about self-fulfilling prophecies. For example, the assumption that a particular technology is fail-safe could cause a relaxation in safety measures and lead a decision maker to avoid the necessary mechanism to circumvent or to prevent the potential problem altogether. In other words, a technological disaster with an occurrence probability perceived to be nonexistent could not be justified as worthy of prevention, such that necessary resources are committed or deployed. Unfortunately, such an assumption might be devastating. For example, technology tends to become more interactively complex, especially when maintenance decreases (Heath, 2002; Olaniran & Williams, 2004; Perrow, 1984). To this end, decision makers set in motion actions that are consistent with their assumptions (Olaniran & Williams, 2001, 2004; Weick, 1988). Thus, the ensuing potential crises from such oversight increase, since the decision maker engaged actions that were only consistent with their assumptions (see Perrow, 1984; Weick, 1988). Similarly, Jaques (2007) suggests that early warning signs and environmental scanning are important in helping to prevent a crisis, but they have no value when management ignores, denies, or tries to suppress them for whatever reason.

Expectation and enactment are intertwined and not mutually exclusive categories in the AMCM. Furthermore, expectation influences enactment and vice versa. However, to say that expectation engages the enactment process is not to make a linear assumption about the two components. Thus, an assumption of interdependency between the two factors is assumed, given that the enactment process could result in, or set into motion, a condition where the expectation is made based on the action taken during the enactment.

While the anticipatory model emphasizes pre-crisis management activities, other components within the model also address the importance of active engagement during crises. This includes the concept of *control*. Control is multifaceted and presented in the model as a way to address how well institutions and decision makers are in command of crisis situations. Control, however, is measured in relative terms and includes stakeholders and relevant environments (e.g., public perception). The notion of control is necessitated and called for in communication interactions where the ability to prepare and to implement effective crisis management programs includes information dissemination and media relations—not to mention the flexibility to make changes when necessary (i.e., rigidity avoidance). Moreover, control is linked with enactment and with expectation. Perceived control regarding a particular crisis event can influence the expectations that decision makers can have about that event, for example, in how much one can counter the factors that feed the potential crisis. Additionally, the enactments that decision makers engage can be informed by the level of control perceived over the crisis situation.

Anticipation as a Vigilant Response

The notions of enactment and expectation are important in the AMCM and in crisis prevention, but they do not constitute an automatic vigilant response to crises. Both expectation and enactment processes are interdependent. Expectations lead to actions that specify where one places decision emphasis (e.g., preventive or post-crisis mode) (Lukasewski, 2013; Olaniran & Scholl, 2014; Sellnow, Ulmer, & Snider, 1998; Zaremba, 2015).

Therefore, the vigilant decision-making process suggests that effective decision makers should go through the decision-making process by carefully analyzing the situation, setting goals, and evaluating outcomes in a relatively efficient manner (Hirokawa & Rost, 1992; Williams & Olaniran, 1994, 2008; Zaremba, 2015). The process centers on careful assessment and understanding of both the external and internal environment, along with communication behaviors within the CCCs. In essence, a balancing act between CCC internal communication strategies and those of community leaders or external personnel and governmental regulations is needed (Olaniran & Williams, 2001; Shrivastava, 1988; Zaremba, 2015). Thus, vigilant decision making centers on both internal and external communication behaviors, which are not mutually exclusive. The internal communication process helps direct organizational activities focusing on the maintenance of stability and structure, while external communication focuses on bringing about change that establishes a decision path for an organization (Anthonissen, 2008; Cheney & Christensen, 2001; Kreps, 1990; Leighton & Shelton, 2008; Olaniran & Williams, 2001).

An important factor to note is that the vigilant decision-making process is best facilitated under a stress-free or reduced-stress environment. Anticipating a crisis focuses on prevention, which implies preparatory practices that can reduce uncertainties and stress when crises eventually occur. Pearson, Kovoor-Misra, Clair, and Mitroff (1997) indicate that practice does not necessarily make for perfection, but it leads to clear thinking and to smoother action, especially when decisions must be made under duress (see also Anthonissen, 2008; Jaques, 2007).

Predicting and anticipating crises would only constitute a vigilant response when the process has managed to scrutinize all of the imaginable potential crises and implement plans for preventing them. This means identifying the *triggering event*, a specific event that is detectable according to place, time, and agents (Anthonissen, 2008; Shrivastava, 1988; Zaremba, 2015). The point is that while triggering events might have a low probability of occurrence, they can, and do, signal the occurrence of a potential catastrophic disaster that must be addressed (Jaques, 2007; Kates, 1977; Lukasewski, 2013; Olaniran & Williams, 2001; Scholl, Williams, & Olaniran, 2005; Pearson, Kovoor-Misra, Clair, & Mitroff , 1997; Pearson & Mitroff, 1993; Udwadia & Mitroff, 1991). Research has found that crises with the most consequences occur because nobody thinks about them (Heath, 2012; Lukasewski, 2013; Mitroff, Harrington, & Gai, 1996; Olaniran, Scholl, Williams, & Boyer, 2012; Pearson, Kovoor-Misra, Clair, & Mitroff, 1997). Thus, events and factors must be scrutinized for places and for conditions where actions or lack of actions can escalate crisis situations exponentially. Technologies, for example, are interactively complex; consequently, by anticipating smaller problems, one can address them before they spread disaster in a community and develop into full-blown crises. Because stress accompanies all crises, the goal is to counteract and ultimately to reduce the stress level in crisis management.

In an effort to facilitate a stress-free or reduced-stress environment for CCC decision makers, attention to two key strategies during the anticipation process are beneficial for enhancing a vigilant response. The two factors involve elements of *rigidity* and *control* introduced or designed by decision makers.

Rigidity: *Rigidity* is the degree of inflexibility built into a particular action or process. Thus, the degree of rigidity exercised by CCC decision makers could determine the degree of vigilance in a crisis response. Weick (1988) illustrates the notion of rigidity by looking at the relationship between enactment and commitment. He indicates that as actions become public and irrevocable, they become difficult to change and that these actions also become more volitional, which in turn makes their explanation more tenacious. This view is consistent with vigilant interaction theory, which argues that the way in which individuals view problems, available options, and consequences will determine the quality of the choices they make (Hirokawa & Rost, 1992). Rigidity, however, is not an independent variable in the AMCM. Rather, rigidity is considered a covariate of expectations that are influenced by the types of assumptions decision makers formulate and, consequently, the ensuing actions they take (i.e., enactment). For instance, at times management turns a blind eye to warning signs or continues with the attitude of business as usual until the situation turns into a full-blown crisis. A case in point is Johnson & Johnson not recalling its Children's Motrin after being made aware that the product did not dissolve properly when taken; Johnson & Johnson became aware of this defect a few years before it became a public crisis (Olaniran, Scholl, Williams, & Boyer, 2012). As a matter of fact, Johnson & Johnson did its best to cover up the recall by having its salespeople quietly go to the stores carrying the defective product and buy back the product from vendors to avoid public detection. The Johnson & Johnson case illustrates the connection between assumptions and resulting actions. When a company assumes that no harm will come to the public for not knowing about a product defect, it might choose to take covert actions to cover up the defect, which in turn might only cause further harm instead of preventing it. Rigidity can be positive or negative. For example, tenacious justifications that involve rigid explanations about certain actions could be advantageous in crisis situations for clarifying ambiguity and confusion (Salancik, 1977, Staw, 1980; Williams & Olaniran, 1998). Furthermore, justification provides structure for maintaining accurate views and for analyzing available options (Lukasewski,

2013; Weick & Sutcliffe, 2007; Zaremba, 2015). On the other hand, rigidity has the potential to be negative, especially when it creates a "blinder effect" that prevents individuals and key decision makers from considering competing alternatives. In such cases, one fails to acknowledge that a decision or an explanation could adversely hinder or intensify crises. For example, community leaders are often convinced that meeting regulatory standards will absolve them of liabilities (Scholl, Williams, & Olaniran, 2005). When this happens, decision makers fail to recognize that death or losses attributable to organizational crises may not be dismissed on the basis of simple adherence to the minimal safety standards. This is because different rules and factors come into play depending on the environment affected. Some scholars have addressed this concern by suggesting that organizational responsibility in crisis situations should be decided on the extent to which a crisis could have been anticipated and prevented (Jaques, 2007; Olaniran & Williams, 2001, 2011; Pearson, Kovoor-Misra, Clair, & Mitroff, 1997; Sethi, 1987; Zaremba, 2015). The following statement illustrates a danger inherent with rigidity: When a person is committed to the view that fluctuations in electricity cause 90% of the variances in electronic gauges, the possibility that a much different percentage is possible will not be entertained until the crisis is at the advanced stage (Weick, 1988, Weick & Sutcliffe, 2007). Similarly, Jaques (2007) presents the fact that some organizations have a crisis plan just for the sake of having a plan, with no intent of ever implementing the plan.

Control: *Control* represents the degree of influence that CCC members have at their disposal to manage crises proactively. Control is elusive because it has to do with perception, especially when the influence is indirect in nature, which is often the case. Control influences crises in the sense that it reflects how individuals, leaders, and decision makers at large respond to crisis situations. For example, when individuals see themselves having the ability and power to do something about a crisis, it is more likely that they will take appropriate action. Thus, ability and authority go hand in hand and establish empowerment, which could extend to a vigilant response in crisis management. While planning for a crisis or a disaster is not an end in effective crisis management, it provides a good starting point to have control over how one handles the crisis.

CCCs can have a hand in educating decision makers, leaders, and the community on how much control they actually have in certain crisis situations. Empowering decision makers and community residents with crisis management programs provides an opportunity for individuals to exercise control over their own behaviors by exploring their personal abilities and, hence, their abilities to avert crises. If people are given the opportunity to handle a variety of issues (expanded roles) based on their expertise or when equipped with the right information, they will be willing to do more than if they are restricted. Empowerment creates the belief among employees or crisis practitioners that they should pay more attention to their surroundings and to important issues (Jaques, 2007; Lukasewski, 2013; Olaniran & Williams, 2001, Olaniran, Williams, & Coombs, 2012; Weick, 1988). Empowerment also leads to greater abilities to control and to respond appropriately to the environment. In the context of CCCs, it is important that individuals who work in the centers are able to exercise their roles with few restrictions. In this light, it is important that the centers are staffed with the right people who have the necessary expertise to perform whatever duties are assigned. The expertise will serve as a buffer that can mitigate the need to go through other gatekeepers before taking appropriate actions.

Furthermore, when CCCs are staffed with the right personnel and they are empowered, individuals are able to respond more quickly to crisis-triggering events as they unfold by taking necessary actions. For example, in the earlier illustration, an employee looking at the high fluctuations in gauges might choose to report the irregularities to the appropriate quarters or to a technician rather than dismiss the variances as simple fluctuations in electricity, thereby preventing a crisis from occurring. By contrast, an individual who has been trained that fluctuations fall within certain thresholds may see some fluctuations as occupying a certain range of acceptance and thus dismiss the fluctuations as anything but a crisis-triggering event that warrants further investigation (Olaniran & Williams, 2001; Perrow, 1984).

Giving CCC members greater control implies that the control might have to be redistributed. Top-level organizational members should be willing to relinquish some of their control to the lower-level members; this could be accomplished by facilitating a cooperative climate and working relationships among members (Kreps, 1990; Quinn & Spreitzer, 1996). Individuals who are closer to crisis-triggering events should be allowed to attend to the situation rather than having to wait for approval from higher-ups (Perrow, 1984; Quarantelli, 1997). This idea counters the notion of having a central spokesperson or centralized decision making, but it does not have to be the case, because the emphasis is on crisis prevention, which may not negate protocols in a crisis response. Furthermore, this process addresses the need for CCC staff and leaders to determine the optimal span of control for particular tasks (Anthonissen, 2008; Clarke & Perrow, 1996; Leighton & Shelton; 2008; Olaniran & Williams, 2001; Quinn

& Spreitzer, 1996). This process stresses decentralization rather than centralized information dissemination that is prevalent in crisis management literature; the two are not necessarily in contradiction. Centralization addresses information and actions geared toward individuals who are external to the organization, such as the public and the media (Shallowitz, 1987). Given the need for consistency, centralization is prudent. However, decentralization of information and action is stressed because of the internal focus of crisis prevention, whereas centralization will only complicate issues by delaying actions due to a greater span of control resulting from different organization constituents and job responsibilities (Tabris, 1984; Zaremba, 2015). To this end, we recommend that a small span of control be used in the anticipatory crisis management model because of the interdependent and nonroutinized nature of tasks (Clarke & Perrow, 1996).

Decentralization also emphasizes the importance of career paths in organizations and in their crisis prevention and management. Weick and Sutcliffe (2007) stressed that decision makers who moved up the organizational ladder via technical ranks usually possess hands-on experience and expertise to deal with triggering events before they become a full-blown crisis by troubleshooting problems in the community environment. The experience could further help decision makers recognize the importance of letting people who are closer to the source of the problem attend to it.

Another implication is the need to structure or to restructure organizational management promotion policies. For example, in a crisis-prone environment, CCCs are best staffed by personnel who have technical expertise to scour the environment, analyze it, and offer results that could eliminate potential problems. More importantly, it would help if CCCs could hang on to these groups of experts on a long-term basis, which would align with the need to maintain consistencies in decision-making types and provide incentives accordingly (Shrivastava & Mitroff, 1987; Weick, 1988). Relinquishing control might not be an easy task at first. However, having personnel with different levels of expertise who identify with the degree and boundary of control may make yielding control a nonissue or, at least, make it a relatively easier goal to accomplish. Furthermore, when the focus is on superordinate goals focused on public good, it can help prevent the tunnel vision that creates crises of management failure—where short-term goals supersede long-term ones (Lerbinger, 2012; Lukasewski, 2013).

More significant is the effect of both rigidity and control on the interaction patterns within CCCs and within communities. Excessive rigidity and inappropriately distributed control tend to result in varieties of faulty decision-making schemes, including (a) failing to recognize positive qualities of alternate decision options; (b) overlooking critical negative aspects of a choice; (c) exaggerating positive aspects of a decision; and (d) overestimating negative consequences of a decision (Hirokawa & Rost, 1992; Olaniran, 1994; Zaremba, 2015). Crisis decision-making processes resulting in one or more of these problems usually hinder decisions and their outcomes, which in a way can render a crisis management procedure a mere "fantasy document" (Clarke & Perrow, 1996; Jacques, 2007). However, setting plans in motion that avoid these faulty decision-making processes can facilitate vigilant decision making (Hirokawa & Rost, 1992; Jaques, 2007; Olaniran & Scholl, 2014).

The AMCM presents a proactive view of crisis management where efforts are made to avert and to eliminate crises altogether. When crises eventually occur, the proactive measures should minimize the eventual negative and devastating consequences. Therefore, the anticipatory model of crisis management is designed to aid in the preparation and analysis of potential crises. Public relations personnel can then better maintain a positive public image for their cities and respective citizens and organizations within their communities. By following the foundations and tenets of the AMCM, CCCs could emerge better prepared to handle crises of all kinds (i.e., natural and unnatural) that threaten their municipalities. They could also serve as avenues for crisis training and for readiness for other communities.

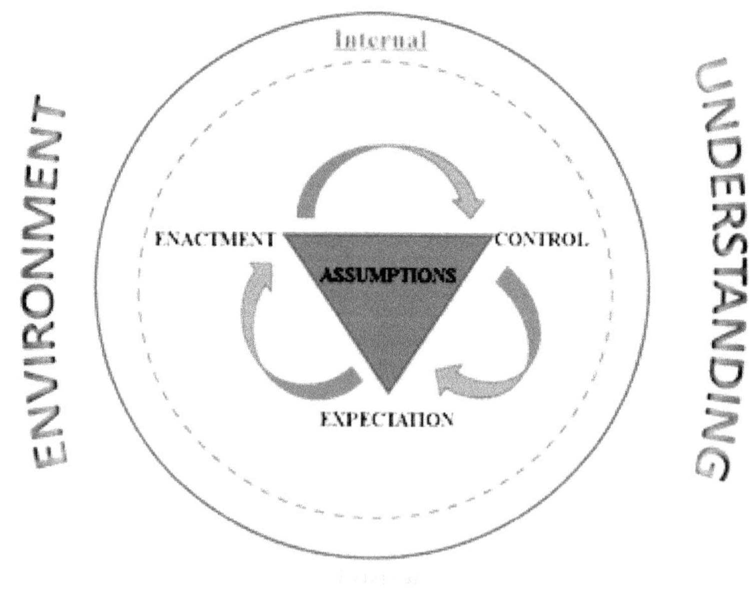

Figure 1: Anticipatory Model of Crisis Management.

Summary of Crisis Communication Centers

With an underpinning in the anticipatory model of crisis management and communication, communities can develop CCCs to determine community-specific risks and the best ways to address these risks in their communities. The assessments of risks would involve investigations similar to crisis audits and to an understanding of the media outlets.

Crisis Preparation

Crisis preparation for newly created CCCs is similar to standard procedures for crisis-management teams (Coombs, 1999; Fink, 1986; Lerbinger, 2012; Olaniran & Williams, 2001; Pauchant & Mitroff, 1992; Scholl, Williams, & Olaniran, 2005). The specific needs of CCCs require early attention to the selection of the CCC team members, completion of a community assessment, identification of spokespersons, and attention to the team communication and action plan.

The CCC team will be a collection of community and municipal members who have knowledge and expertise, represent the community, and also fulfill specific needs of the team. It is for this reason that we recommend housing a CCC on a university campus or in a learning institution, given that individuals, scientists, communication scholars, and program planners are readily available and possess the necessary expertise for making such a center effective. Ideally, CCCs will include individuals with a working knowledge of elements in the community—industrial, technical, and agricultural—that is potentially crisis imbued or that could become targets for terror attacks. The CCC team will also need individuals with communication and media expertise, representatives of local government, and at least one individual who can successfully coordinate the operations of the team. The size of the CCC team will likely fluctuate depending on the size of the community and the number of potential crises and terrorism targets. Crisis team members must have the ability to make group decisions, function as a team, enact a crisis plan, and listen to each other and to members of the greater community (Coombs, 1998; Coombs & Chandler, 1996; Jaques, 2007).

In addition to working as a team, CCC members must possess the expert knowledge that allows them to assess crisis risk independently of the team (Olaniran & Williams, 2004). In particular, completing a community assessment is the first critical project for the newly established CCC team. Team members should attend to generally recognized threat targets that are characteristic of that community. However, in order to do this efficiently and effectively, members' localized knowledge and expertise are crucial. Notwithstanding, CCC team members need to move beyond the more obvious terrorist targets of national landmarks and nuclear power plants and, rather, identify potential sites of interests that may be obscure but that may be nonetheless vulnerable. For example, the innocent combination of the purchase of explosive material or the rental of a moving van in one community could be viewed as a weapon in another location. A case in point was the ability to foil the terror plot by a chemical engineering student at Texas Tech University. The student ordered explosive chemicals online as part of a plan to hide bomb materials inside dolls and baby carriages in order to blow up dams, nuclear plants, or the Dallas home of former president George W. Bush (Goldman & Blaney, 2011). In a case such as this, the ability to put together pieces of information that would otherwise pose no threat independently can foretell disaster and is the key role of CCCs.

Crisis preparation is among the first of many stages of equipping a community for handling potential crises and for establishing a crisis communication center that is appropriate in size and scope. Throughout this book, we propose that communities across the country create CCCs that would be composed of municipal and community members who can fulfill specific functions on a team tasked with preparing for a crisis and responding to a crisis's aftermath. The purpose of this chapter is to introduce the concept of the CCC and to provide a justification for its incorporation into any community, large or small. The CCC is framed around the AMCM, which illustrates how decision makers, leaders, practitioners, and community members can utilize information and organize local communication networks to be vigilant for crisis markers and to allocate the necessary resources for handling crisis events or threats.

The rest of this book explores the research- and practice-oriented issues related to community-based crisis management. Chapters 2 to 5 address research topics associated with crisis management and how they apply to the development of local CCCs. Chapter 2 addresses the role the CCC plays in the community and discusses the types of information available to help researchers and practitioners make accurate assessments of a community's vulnerabilities and resources. Chapter 3 explores the nature of stakeholders and their roles in crisis preparation and management, in particular, how stakeholders are essential to the CCC and how they assist in the implementation of the AMCM. Chapter 4 addresses the role of technology and mass media in crisis communication and its interaction with community-based crisis efforts. Part I of the book concludes with Chapter 5, which discusses the relevant research on community-based crisis communication and management and implications for future research.

The second half of the book, Part II, contains chapters that provide more specific guidelines for designing, implementing, and maintaining a CCC within a community. Chapter 6 presents an overview of the common types of biologic and health-related crises and the health impacts of organizational misconduct. The chapter then outlines specific strategies that can help address the most common health disasters, such as improved infrastructure and vaccination campaigns, and how CCCs can help design and implement these strategies. Chapter 7 discusses cultural diversity and racial tensions within a community, and talks about the CCC's role in appealing to diverse audiences and in constructing mediated messages that are culturally sensitive. Chapter 8 reviews various ethical perspectives and philosophies and applies them to the operations of the CCC. The chapter also reviews common ethical dilemmas unique to crisis situations and gives prescriptions for how CCC personnel can handle them. Chapter 9 makes the case for conducting drills and for practicing crisis management protocols developed by a CCC. The essential components of a plan are presented along with factors that influence its adaptability to various communities. Finally, we present Chapter 10 as a guide for professionals in public relations, medicine, law enforcement, and public office, and other individuals who could serve as CCC personnel. Included in this chapter are sample assessment tools, message templates, and procedural guidelines to assist in the creation and maintenance of a CCC.

References

Anthonissen, P. (2008). *Crisis communication: Practical PR strategies for reputation management & company survival*. Philadelphia: Kogan Page.

Burke, K. (1961). *Attitudes toward history*. Boston: Beacon.

Cheney, G., & Christensen, L. T. (2001). Identity at issue: Linkages between "internal" and "external" organizational communication. In F. M. Jablin & L. L. Putnam (Eds.), *The new handbook of organizational communication* (231–269). Thousand Oaks, CA: Sage.

Clarke, L., & Perrow, C. (1996). Prosaic organizational failure. *American Behavioral Scientist*, 39, 1040–56.

Coombs, W. T. (1998). An analytic framework for crisis situations: Better responses from a better understanding of the situation. *Journal of Public Relations Research, 10*(3), 177–91.

Coombs, W. T. (1999). *Ongoing crisis communication: Planning, managing, and responding*. Thousand Oaks, CA: Sage.

Coombs, W. T., & Chandler, R. C. (1996). Crisis teams: Revisiting their selection and training. In L. Barton (Ed.), *New avenues in risk and crisis management* (pp. 7–15). Las Vegas: University of Nevada, Las Vegas Small Business Development Center.

European Commission (2015). Humanitarian aid and civil protection. Retrieved July 23, 2015, from http://ec.europa.eu/echo/what/civil-protection/emergency-response-coordination-centre-ercc_en

Fink, S. (1986). *Crisis management: Planning for the inevitable*. New York: AMACOM.

Goldman, A., & Blaney, B. (2011, February 24). Texas student held on terrorism charges. http://www.statesman.com/news/news/state-regional/texas-digest-trial-date-set-for-student-held-in-bo/nRYh2/.

Gregory, A. (2005). Communication dimensions of the UK foot and mouth disease crisis 2001. *Journal of Public Affairs*, 5, 312–28.

Heath, R. L. (2002). Issues management: Its past, present and future. *Journal of Public Affairs, 2*(4), 209–14.

Heath, R. L., & Sultan, J. (2012). Pre-crisis management and communication: Slippery steps or solid footing. In B. Olaniran, D. Williams, & W. Coombs (Eds.), *Pre-crisis planning, communication, and management: Preparing for the inevitable* (pp. 101–24). New York: Peter Lang.

Hirokawa, R. Y., & Rost, K. (1992). Effective group decision making in organizations. *Management Communication Quarterly*, 5, 267–88.

International Association for Suicide Prevention (2015). Resources: Crisis centers. Retrieved July 22, 2015, from http://www.iasp.info/resources/Crisis Centers/.

Jaques, T. (2006). Issue management: Process versus progress. *Journal of Public Affairs*, 6(1), 69–74.

Jaques, T. (2007). Issue management and crisis management: An integrated, non-linear, relational construct. *Public Relations Review, 33*(2), 147–57.

Jaques, T. (2012). Issue management as strategic aspect of crisis prevention. In. B. Olaniran, D. Williams, & T. Coombs (Eds.), *Pre-crisis planning, communication, and management: Preparing for the inevitable* (pp. 17–36). New York: Peter Lang.

Kates, R. W. (1977). *Managing technological hazard: Research needs and opportunities*. Boulder: University of Colorado Institute of Behavioral Science.

Kilduff, M. (1996). Making sense of sensemaking: Into the jungle with Karl Weick. *Journal of Management Inquiry*, 5, 246–49.

Kreps, G. L. (1990). *Organizational communication: Theory and practice* (2nd ed.). New York: Longman.

Leighton, N., & Shelton, T. (2008). *Crisis communication, Practical PR strategies for reputation management and company survival*. Philadelphia: Kogan Page.

Lerbinger, O. (2012). *The crisis manager* (2nd ed.). New York: Routledge.

Lindell, M. K., & Perry, R. W. (2004). *Communicating environmental risk in multiethnic communities*. Thousand Oaks, CA: Sage.

Lukasewski, J. E. (2013). *On crisis communication: What your CEO needs to know about reputation, risk and crisis management*. Brookfield, CT: Rothstein.

Mitroff, I. I., Harrington, K., & Gai, E. (1996, September). Thinking about the unthinkable. *Across the Board, 33*(8), 44–48.

Mitroff, I. I., Pauchant, T., & Shrivastava, P. (1989). Can your company handle a crisis? *Business & Health, 7*(5), 41–45.

Olaniran, B. A. (1994). Group performance and computer-mediated communication. *Management Communication Quarterly, 7*, 256–81.

Olaniran, B. A. (2007). The role of perception in crisis management: A tale of two hurricanes. *Multicultural Education, 15*(2), 13–16.

Olaniran, B. A., & Scholl, J. C. (2014). New England compounding center meningitis outbreak: A compounding public health crisis. *Journal of Risk Analysis and Crisis Response, 4*(1), 34–42.

Olaniran, B. A., Scholl, J. C., Williams, D. E., & Boyer, L. (2012). Johnson and Johnson phantom recall: A fall from grace or a re-visit of the ghost of the past. *Public Relations Review, 38*, 153–55.

Olaniran, B. A., & Williams, D. E. (2001). Anticipatory model of crisis management: A vigilant response to technological crises. In R. L. Heath (Ed.), *Handbook of public relations* (pp. 487–500). Thousand Oaks, CA: Sage.

Olaniran, B. A., & Williams, D. E. (2004). Burkian counternature and the vigilant response: An anticipatory model of crisis management and technology. In D. Millar & R. Heath (Eds.), *Responding to crisis: A rhetorical approach to crisis communication* (pp. 75–94). Mahwah, NJ: Lawrence Erlbaum.

Olaniran, B. A., & Williams, D. E. (2008). Applying anticipatory and relational perspectives to the Nigerian delta region oil crisis. *Public Relations Review, 34*, 57–59.

Olaniran, B. A., Williams, D. E., & Coombs, W. T. (Eds.). (2012). *Pre-crisis planning, communication, and management: Planning for the inevitable*. New York: Peter Lang.

Pauchant, T. C, & Mitroff, I. I. (1992). *Transforming the crisis-prone organization: Preventing individual, organizational, and environmental tragedies*. San Francisco: Josey-Bass.

Pearson, C. M., & Clair, J. A. (1998). Reframing crisis management. *Academy of Management Review, 32*(1), 59–76.

Pearson, C. M., Kovoor-Misra, S., Clair, J., & Mitroff, I. I. (1997). Managing the unthinkable. *Organizational Dynamics, 26*(2), 51–64.

Pearson, C. M., & Mitroff, I. I. (1993). From crisis prone to crisis prepared: A framework for crisis management. *The Executive, 7*, 48–59.

Perrow, C. (1984). *Normal accidents: Living with high risk technologies*. New York: Basic Books.

Quarantelli, E. L. (1997). Ten criteria for evaluating the management of community disasters. *Disasters, 21*(1), 39–56.

Quinn, R. E., & Spreitzer, G. M. (1996). Seven questions every leader should consider. *Organizational Dynamics, 26*, 37–50.

Rape Crisis Network Europe (n.d.). Help is there for you, no matter where you are. Retrieved July 22, 2015, from http://www.rcne.com/.

Salancik, G. R. (1977). Commitment and the control of organizational behavior and belief. In B. M. Staw & G. R. Slancik (Eds.), *New directions in organizational behavior* (pp. 1–54). Chicago: St. Clair.

Scholl, J. C., Williams, D. E., & Olaniran, B. A. (2005). Preparing for terrorism: A rationale for the crisis communication center. In D. O'Hair, R. Heath, & G. Ledlow (Eds.), *Community preparedness and response to terrorism: Communication and the media* (pp. 243–68). Westport, CT: Praeger.

Sellnow, T. L., Ulmer, R. R., & Snider, M. (1998). The compatibility of corrective action in organizational crisis communication. *Communication Quarterly*, *46*, 60–74.

Sethi, S. P. (1987). Inhuman errors and industrial crisis. *Columbia Journal of World Business*, *22*, 101–7.

Shallowitz, D. (1987, April 13). Preparation needed to handle catastrophes. *Business Insurance*, 30.

Shrivastava, P. (1987). *Bhopal: Anatomy of crisis*. Cambridge, MA: Ballinger.

Shrivastava, P. (1988), Industrial crisis management: Learning from organizational failures. *Journal of Management Studies*, *25*(4), 283–84.

Shrivastava, P., & Mitroff I. I. (1987). Strategic management of corporate crisis. *Columbia Journal of World Business*, *22*, 5–11.

Shrivastava, P., Mitroff, I. I., Miller, D., & Miglani, A. (1988). Understanding industrial crises. *Journal of Management Study*, *25*, 285–303.

Smircich, L., & Stubbart, C. (1985). Strategic management in an enacted world. *Academy of Management Review*, *10*, 724–36.

Staw, B. (1980). Rationality and justification in organizational life. In L. Cummings & B. Staw (Eds.), *Research in organizational behavior* (pp. 45–80). Greenwich, CT: JAI.

Tabris, M. D. (1984). Crisis management. In *Experts in action: Inside public relations*. New York: Longman.

Udwadia, F. E., & Mitroff, I. I. (1991). Crisis management and the organizational mind: Multiple models for crisis management from field data. *Technological Forecast and Social Change*, *40*, 33–52.

Weick, K. E. (1988). Enacted sensemaking in crisis situations. *Journal of Management Studies*, *25*, 305–17.

Weick, K. E. & Sutcliffe, K. (2007). *Managing the unexpected: Assuring high performance in an age of complexity* (2nd ed.). New York: Jossey-Bass.

Williams, D. E., & Olaniran, B. A. (1994). Exxon's decision-making flaws: The hypervigilant response to the *Valdez* grounding. *Public Relations Review*, *20*, 5–18.

Williams, D. E., & Olaniran, B. A. (1998). Expanding the crisis planning function: Introducing elements of risk communication. *Public Relations Review*, *24*(3), 387–400.

Zaremba, A. J. (2015). *Crisis communication: Theory and practice*. New York: Routledge.

Additional Resources:

http://news.msn.com/us/suspicious-powder-mailed-to-hotels-near-super-bowl

Case Study

Ebola & CDC: Cases of Ebola Diagnosed in the United States

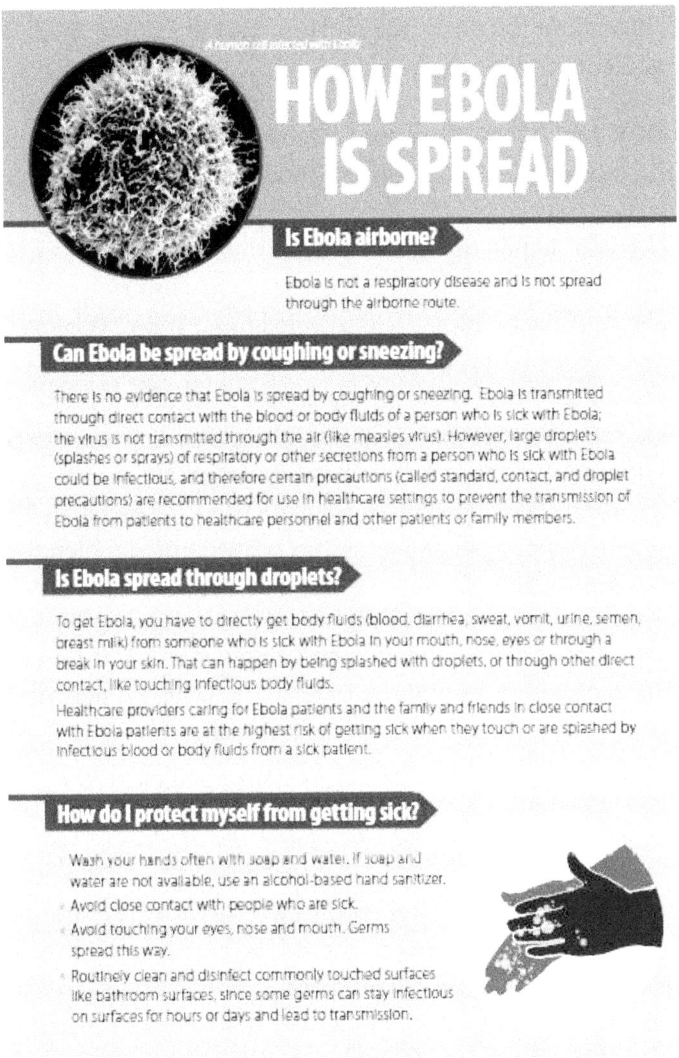

HOW EBOLA IS SPREAD

Is Ebola airborne?

Ebola is not a respiratory disease and is not spread through the airborne route.

Can Ebola be spread by coughing or sneezing?

There is no evidence that Ebola is spread by coughing or sneezing. Ebola is transmitted through direct contact with the blood or body fluids of a person who is sick with Ebola; the virus is not transmitted through the air (like measles virus). However, large droplets (splashes or sprays) of respiratory or other secretions from a person who is sick with Ebola could be infectious, and therefore certain precautions (called standard, contact, and droplet precautions) are recommended for use in healthcare settings to prevent the transmission of Ebola from patients to healthcare personnel and other patients or family members.

Is Ebola spread through droplets?

To get Ebola, you have to directly get body fluids (blood, diarrhea, sweat, vomit, urine, semen, breast milk) from someone who is sick with Ebola in your mouth, nose, eyes or through a break in your skin. That can happen by being splashed with droplets, or through other direct contact, like touching infectious body fluids.

Healthcare providers caring for Ebola patients and the family and friends in close contact with Ebola patients are at the highest risk of getting sick when they touch or are splashed by infectious blood or body fluids from a sick patient.

How do I protect myself from getting sick?

- Wash your hands often with soap and water. If soap and water are not available, use an alcohol-based hand sanitizer.
- Avoid close contact with people who are sick.
- Avoid touching your eyes, nose and mouth. Germs spread this way.
- Routinely clean and disinfect commonly touched surfaces like bathroom surfaces, since some germs can stay infectious on surfaces for hours or days and lead to transmission.

(http://www.cdc.gov/vhf/ebola/pdf/infections-spread-by-air-or-droplets.pdf)

How Ebola is Spread[PDF - 1 page](http://www.cdc.gov/vhf/ebola/pdf/infections-spread-by-air-or-droplets.pdf)

- **October 23, 2014**—The New York City Department of Health and Mental Hygiene reported a case of Ebola in a medical aid worker who had returned to New York City from Guinea, where the medical aid worker had served with Doctors without Borders.
  - The diagnosis was confirmed by CDC on October 24.
  - The patient has recovered and was discharged from Bellevue Hospital Center on November 11.

- **October 15, 2014**—A second healthcare worker who provided care for the index patient at Texas Presbyterian Hospital, tested positive for Ebola.
  - This second healthcare worker was transferred to Emory Hospital in Atlanta, Georgia.
  - The healthcare worker had traveled by air from Dallas to Cleveland on October 10, and from Cleveland to Dallas on October 13. CDC worked to ensure that all passengers and crew on the two flights were contacted by public health professionals to answer their questions and to arrange follow-up as necessary.
  - The patient has since recovered and was discharged on October 28.
  - By November 3, all passengers on both flights completed the 21-day monitoring period.

- **October 10, 2014**—A healthcare worker at Texas Presbyterian Hospital who provided care for the index patient, tested positive for Ebola.
  - The healthcare worker was isolated after the initial report of a fever and subsequently moved to the National Institutes for Health (NIH) Clinical Center.
  - The patient has since recovered and was discharged on October 24.

- **September 30, 2014**—CDC confirmed the first laboratory-confirmed case of Ebola to be diagnosed in the United States in a man who had traveled to Dallas, Texas, from Liberia.
  - The man did not have symptoms when leaving Liberia, but developed symptoms approximately four days after arriving in the United States.
  - The man sought medical care at Texas Presbyterian Hospital of Dallas after developing symptoms consistent with Ebola. Based on his travel history and symptoms, CDC recommended testing for

Ebola. The medical facility isolated the patient (i.e., index patient) and sent specimens for testing at CDC and at a Texas laboratory.
- Local public health officials identified all close contacts of the index patient for daily monitoring for 21 days after exposure.
- The patient passed away on October 8.
- By November 7, all contacts of the patient completed the 21-day monitoring period.

CDC recognizes that any case of Ebola diagnosed in the United States raises concerns, and any death is too many. Medical and public health professionals across the country have been preparing to respond to the possibility of additional cases. CDC and public health officials in Texas, Ohio, and New York took precautions to identify people who had close personal contact with the patients, and healthcare professionals have been reminded to use meticulous infection control at all times.

From http://www.cdc.gov/vhf/ebola/outbreaks/2014-west-africa/united-states-imported-case.html

Discussion Questions:

1. How is Ebola spread?
2. Can a local community such as Dallas or elsewhere take comfort in the fact that because a disease such as Ebola happens in a remote area of the world it would never happen in their communities?
3. How can a community plan and prepare for a disease outbreak such as Ebola using the AMCM theory? Provide practical application(s).
   For a version of the chronicle of events, see *Dallas Morning News* article: http://www.dallasnews.com/ebola/headlines/20141206-er-doctor-discusses-role-in-ebola-patients-initial-misdiagnosis.ece

· 2 ·

CCC AND THE MULTIPLE STAKEHOLDERS

Chapter Highlights

- "Stakeholder" defined
- The role of stakeholders in crisis management
- Multiple stakeholders and their different needs

Introduction

Stakeholders are people who have a vested interest in a particular crisis or disaster. Zaremba (2015) defines or characterizes "stakeholders" as audiences who receive messages during crises. However, during crises or disasters, there are multiple stakeholders. For instance, stakeholders may have direct interest in the crisis and others may not, but affected organizations may see them as an asset in managing or in responding to a crisis. Specifically, Zaremba refers to the latter group as individuals who receive messages and those that organizations want to have a stake in the crisis. Therefore, stakeholders are people who can be an extension of resources available to organizations or can be victims affected by crises or disaster events. This chapter explores the nature of stakeholders and their roles in crisis preparation and management, in particular, how stakeholders are essential to CCCs and how they assist in the implementation of the AMCM.

Crises do not occur in a vacuum; rather, their occurrence impacts people and affects their lives, and in certain situations it often changes people's lives forever. Because a crisis leaves such an indelible impression on people, it is important that pre- and post-crisis management takes into consideration people whose lives are impacted directly or indirectly. Therefore, drawing from the AMCM principles, an organization such as the CCC must understand that a given crisis would impact the stakeholders in their attempt to plan, to prepare, and to gain some level of control over the crisis or disaster. Stakeholders consist of groups or individuals and members of the public affected or those, in turn, who can affect an organization's performance and goal attainment (Bland 1998; Hearit, 2006; Lerbinger, 1997, 2012; Lukaszewski, 2013; Ray, 1999, Stephens, Malone, & Bailey, 2005). Stakeholders are individuals who tend to have the most invested interest in seeing successful crisis communication plans. Organizations, for the most part, function in unstable environments where they need to respond to multiple stakeholders. Even though a CCC is not a for-profit or traditional organization, it is an organization nonetheless, because it must deal with different or multiple stakeholders within its community. Given that a CCC is charged with crisis planning and with disaster management, it is evident that any given crisis or disaster will affect different and multiple stakeholders, and, therefore, CCCs must be able to identify and to put in place a plan for addressing their idiosyncratic needs.

Distinction in Organizational Types and Crisis Disposition

The idea of stakeholders can be linked to the notion of publics. However, the concept of *publics* and *organizations* has been treated in the public relations literature as interchangeable (Leitch & Motion, 2010, p. 100). The interchangeability of the two terms is considered to be no problem, because an organization can serve as a public for another organization (Leitch & Motion, 2010). Nonetheless, Coombs (1993) suggests that the lack of power from most public relations theory makes both terms equivalent. Regardless of the distinction, or lack thereof, between public and organization, it stands to reason that there are multiple publics that public relations and crisis management practitioners need to be concerned with. This notion of multiple publics, or stakeholders, has been reinforced in the work of scholars, who have argued succinctly that there is no general public in public relations—instead, there are fragmented publics (e.g., Leitch & Motion, 2010; Rawlings & Bowen, 2005). Be that as it may, when one is concerned with a CCC, one needs to realize that implications of for-profit and certain traditional organizations may not apply. After all, one is dealing with communities working to address the demands of different community members, such that the term *publics* may be more appropriate for the stakeholders, however diverse or fragmented these publics may be.

Furthermore, there are not as many published cases that demonstrate crisis management among government agencies as there are for traditional or for-profit corporate organizations. Nevertheless, there are cases where some organizations are more prone to crises than others. For example, oil and gas corporations, such as Exxon and BP, have demonstrated a frequency of explosions and oil spills, among others. Like for-profit organizations, it can be argued that certain communities are more prone to certain types of disasters and crises than others. For example, communities located near the Gulf Coast are more likely to experience hurricanes and flooding than others. In essence, a CCC's role is to identify and to acknowledge the role of environments on its community and hence to determine how to meet the needs of different stakeholders in its crisis planning accordingly.

From another standpoint, the nature of publics in a community poses different implications for CCCs than those of corporate organizations. For instance, Avery and Lariscy (2010) argued that publics, patrons, or shareholders of a corporate organization may express their frustration in a given crisis by selling stocks or by refusing to patronize a company again. However, residents of a community may see themselves without such an option, leaving them with the perception that they are stuck with few or no options. Thus, they can call for resignation of the community leaders or elected officials, but that may not help resolve their most pressing and immediate needs. Consequently, CCCs must have the expectation that its publics will expect it to be expert or knowledgeable at dealing or managing different crises and disasters. As a result, publics or stakeholders within their respective communities may be more vulnerable than other publics in terms of decisions made on their behalf by agencies such as the CCC. Consequently, the CCC needs to know its community and its stakeholders and also understand the expectations about crisis and crisis management in order to effectively address their needs (e.g., communication or otherwise).

CCC Roles with Stakeholders

As we alluded to earlier, the role of the CCC is to prevent a crisis from occurring and to minimize the potential harm a crisis may bring if and when it happens. Yet, the CCC, unlike corporate organizations, must be built with a mind-set that any of its efforts in successfully preventing or in alleviating crisis impacts depends on how the stakeholders perceive its actions and inactions (i.e., communication, intent, goals, and effectiveness). For instance, no amount of effort by CCC members would accomplish these goals if members of the public fail to pay attention to the CCC's messages regarding risk communication and compliance. In essence, it is important that relations between a community's CCC and its multiple and diverse publics need to be established, nurtured, and restructured, especially ahead of a crisis. This is in accordance with the core tenet of the AMCM model (Olaniran & Williams, 2001). It is from this standpoint that the notion of *dialogue* becomes valuable. Dialogue points to the quality of relationships between people (de Bussy, 2010). Genuine dialogue should focus on establishing a mutual relationship that is not built on the notion of *us* versus *them* or *I* and *thou*. Rather, it is established on the premises of "we are in this together." After all, relationship and relationship management are never static but are, rather, in a state of negotiation and renegotiation (Gilpin & Murphy, 2008). At the same time, de Bussy (2010) identified three types of dialogues taken from Buber (2002), namely, *genuine dialogue*—which focuses on mutual relations and respect; *technical dialogue*—which focuses on objective understanding; and *monologue*—which is considered a disguised dialogue because of the emphasis on an attempt to score points. From these types of dialogues it is easy to see that there is no place for the third type of dialogue (monologue) in crisis management and in particular with CCCs. All stakeholders must be able to count on their community and respective community leaders via the CCC having their best interest in mind, and not have to second-guess them. Unfortunately, trust does not occur naturally; it requires mutual trust, which often takes time to be developed and nurtured, and it is also difficult to regain when lost (Gilpin & Murphy, 2008; Williams & Olaniran, 1998). It is also important to note that the CCC does not allow internal issues or problems to negatively impact the external environment and, subsequently, stakeholders' perception of trust in the organization (Gilpin & Murphy, 2008; Millar & Beck, 2004).

Trust is essential in building and in maintaining relationships with stakeholders. Open and clear communication plays a key role in this area, especially when dealing with multiple stakeholders. For example, the Federal Emergency Management Agency (FEMA) was vilified for its poor handling of Hurricane Katrina in 2005. In 2007, FEMA had the chance to redeem itself in the aftermath of Katrina when charged with the responsibility of handling or helping victims of the California wildfire. It failed yet again. To provide context, in October 2007, there was a wildfire in Southern California that destroyed about half a million acres of land and more than 1,600 homes (Avery & Lariscy, 2011). Along with property destruction, 85 people were injured, 61 of them firefighters, and 14 others lost their lives due to the fires that occurred after an evacuation (Avery & Lariscy, 2011; Phillips, Chernoff, Oppenheim, & Jeras, 2007). FEMA, in its attempt to communicate to the public or to multiple stakeholders that it had control of the situation, decided to hold a press conference with the goal of assuring people that this was a new and competent FEMA. The press conference, however, was considered staged, in that the media were not present and the only individuals able to ask questions were FEMA's own employees, acting as if they were reporters. As a matter of fact, it was reported that FEMA only gave reporters 15 minutes' notice of the briefing, thus making it unlikely for most media personnel to attend (Avery & Lariscy, 2011). Furthermore, reporters or media personnel were given an 800 number to call, but it was a "listen only" line, and the notice indicated that no questions were allowed. The briefing was broadcast live on Fox News, MSNBC, and on other selected outlets, which interrupted their regular programming (Avery & Lariscy, 2011). The backlash from FEMA's deliberate attempt to deceive the public was so bad that even the then secretary of homeland security (Michael Chertoff) called the press conference "one of the dumbest and most inappropriate things I've ever seen since I've been in the government" (Meserve, 2007, as cited in Avery & Lariscy, 2011, p. 321). It is clear from this example how FEMA's action failed to convey trust with stakeholders. Perhaps it can be argued that FEMA's action only reinforced the public's lack of trust in government. Certainly, FEMA's action did little to address the immediate needs of the victims of the wildfire. Specifically, it missed the mark of the presidential directive, which calls for FEMA to "identify, mobilize, and provide at its discretion equipment and resources necessary to alleviate the impacts of the emergency" (Avery & Lariscy, 2011, p. 321). When dealing with stakeholders, it is important that the CCC, as a representative of the community, work to solidify its legitimacy. McVea and Freeman (2005) suggest the need to rethink the idea of stakeholders as real people with needs and with names and faces. Thus, operationalizing dialogue is believed to be based on three useful premises, including listening, positive regards, and willingness to change (de Bussy, 2010; Isaacs, 1999; Senge, 2006).

Multiple Stakeholders and Their Different Demands

With crises and disasters, organizations often have to deal with multiple stakeholders, not just one particular group. These include those directly and indirectly impacted by a given crisis or disaster. The CCC is no exception, and it needs to attend to the different demands of multiple and diverse stakeholders within the community in the sense that the nature of the relationship between the CCC and stakeholders is crucial to how stakeholders perceive the CCC and its credibility. Furthermore, understanding the different needs of multiple stakeholders is instrumental to an organization in assessing how to respond to idiosyncratic needs existing among different groups of stakeholders. At the same time, it has been stressed that organizations' environments are complex and unstable to the extent that internal and external stakeholders cross over boundaries during crises (Gilpin & Murphy, 2008; Lerbinger, 1997; Stephens et al., 2005). Similarly, institutions and organizations, at times, make positive assumptions about certain key stakeholders that may not be accurate. Also, it is not uncommon for key stakeholders to abandon an organization during a crisis to protect their own interests or to distance themselves from a given crisis (Pearson & Mitroff, 1993; Stephens et al., 2005). The shift in allegiance is often characterized as "stakeholder salience" (Mitchell, Agle, & Wood, 1997). Shifting stakeholder relationships create problems for crisis practitioners because the criteria and expectations may be contradictory. Thus, when faced with conflicting demands, crisis managers have to prioritize concerning which groups to focus on and which ones to ignore (Pfeffer & Salancik, 1978). We, however, suggest the need for crisis managers to find an approach that benefits all stakeholders, to the best of their ability. Depending on the context or situation, there is a need to identify all of the different stakeholders involved, given that a crisis can expand the number of salient stakeholders (Ulmer & Sellnow, 2000). For example, after the Boston Marathon bombing in April 2013, the FBI became a stakeholder because it was an act of terrorism. Then there were the Boston publics and Boston Police Department, the marathon participants, and the nation as a whole that must deal with yet another act of terrorism. Therefore, the goal of the CCC in building and in sustaining trust in crisis and disaster situations is to emphasize what are known as "dialogic principles." Dialogic principles contend that an ongoing communication relationship must exist and with a structure that enhances spontaneous and unscripted exchanges (Gilpin & Murphy, 2008), in other words, an opportunity to maintain a two-way communication or forum where different stakeholders are able to ask questions that address their respective needs or that co-create shared meaning.

Pre-Crisis Emphasis

In staying with the intent of pre-crisis rather than post-crisis management for the CCC, it is essential that the emphasis, along with resource allocation, be committed to building relationships and communication interactions with identified stakeholders prior to crisis occurrence. The shift from post-crisis to pre-crisis management is based on the fact that successful crisis management has been found to entail more than *damage control* or getting good publicity out of a bad situation (Olaniran & Williams, 2001; Ulmer & Sellnow, 2000). Furthermore, the CCC's primary responsibility is the safety of the community and its members/citizens. It is not about an organization trying to promote products or services or outlasting a competitor; rather, it is more about making a community safe for all of its residents and going about it in a manner that helps put into good use the collective intelligence of the participating citizens. At the same time, emphasis on the pre-crisis stage leads to crisis preparedness, which ultimately presents a global approach to crisis management by focusing on anticipation, planning, and preparation for a crisis way before it occurs. Thus, engaging multiple stakeholders in the process has the tendency to empower them in knowing that they are part of the CCC decision-making process and not just receivers of communication messages that may be seen as failing to take into account their particular needs.

Given that a crisis creates a high level of uncertainty (Lerbinger, 1997), and uncertainty has been linked to the severity of a crisis and outcomes (Ray, 1999), it is apparent that waiting until a crisis occurs before developing a plan is not the best way to handle or to manage a crisis. Therefore, anticipating a crisis and planning for a crisis through speculation of different scenarios and putting in place a plan for addressing ensuing uncertainties present a more competent way to address a crisis, while at the same time dramatically reducing uncertainties. Mitroff (2004) indicated that there are many uncertainties (e.g., technical and ethical) at the inception of a crisis that challenge the appropriate cause of action. Thus, there should be strategic communication geared toward reducing uncertainties and concerns (Scholl, Williams, & Olaniran, 2005; Stephens, Barrett, & Mahometa, 2013). The strategic communication may also help in putting to rest, or at least address, the initial concerns about whom to blame, response perception, resolution, or mitigation strategies (Ray, 1999; Stephens et al., 2013). In developing a good understanding of stakeholders and their roles in CCCs, it is helpful to discuss stakeholder theory and types of stakeholders. The next section will examine the stakeholder theory of crisis management.

Stakeholder Theory and Crisis

Stakeholder theory (Freeman, 1984) was developed to address how groups and individuals affect an organization and the managerial actions in response to those groups and individuals (Frooman, 1999). As we stated earlier, stakeholders include individuals, groups, or members of the public affected by an organization operation and, specifically, a crisis (Ray, 1999; Stephens et al., 2013). Given the interdependent relationship between organizations and their publics, stakeholders are also individuals and groups who can impact or alter an organization's cause of actions and goals (Bland, 1998; Lerbinger, 1997). The behavior of organizations and their relationship with publics is important because of the mutual effects they have on one another. Organizations and publics do not only have to compete for scarce resources, but they also have limited resources to respond to external groups (Stephens et al., 2013). Therefore, by examining the attributes of stakeholder groups, one can determine how organizations react or interact within their environments and especially with multiple stakeholders within those environments. Other scholars have emphasized the importance of stakeholder relationships and crisis response strategies (e.g., Coombs & Holladay, 2008; Wong & Hung, 2008), and stakeholder reactions to crises (McDonald, Sparks, & Glendon, 2010; Tazin & Kaur, 2014). Coombs (2014), in particular, categorized crisis communication into two types: crisis knowledge management and stakeholder reaction management. Hence, the role of stakeholders in a crisis situation cannot be overemphasized. At the same time, one must understand that the urgency of the need to provide immediate response to a crisis may negatively affect an organization's ability to access all stakeholders. Similarly, it has been suggested that crises create concern for affected organizations in terms of the fear of a bad portrayal by the media or guilt by association, which can affect stakeholder support (Stephens et al., 2013). All of these factors complicate crisis management and produce uncertainties about addressing stakeholders' needs and, consequently, the effectiveness of crisis communication.

Although a crisis presents challenges for an organization and for its respective stakeholders, it can also provide the opportunity for affected organizations to demonstrate communication skills that resolve the crisis in a positive way. It is from this perspective that stakeholder theory perhaps makes its most significant contribution. It recognizes that communication following a crisis can make or break crisis management. Even so, the AMCM focuses on issues management; hence, it is imperative that the CCC identify and engage communication with stakeholders during its crisis planning effort. For instance, one way might be to have a community forum and drill exercises that are open to feedback from stakeholders in an attempt to address any potential problem or challenge that may ensue later on. In particular, the emphasis on the planning function and communication relationship makes for an effective crisis management, especially when stakeholder theory is combined with the AMCM. The rationale is that post-crisis communication cannot occur successfully without proper and adequate pre-crisis communication. It is from this standpoint that scholars have argued the need for organizations to put in place, or to develop, strong pre-crisis relationships with their stakeholders (Heath, 1997; Olaniran & Williams, 2001; Olaniran, Scholl, Williams, & Boyer, 2012; Ulmer & Sellnow, 2000. Heath (1997), in particular, argued for the need for organizations to put in place good standards of corporate responsibility to meet stakeholders' expectations prior to a crisis.

Crisis Communication with Stakeholders

Three types of communication between the CCC and the stakeholders pertain: pre-crisis, crisis, and post-crisis. Although the three appear independent of one another, they are interdependent nonetheless.

Pre-crisis communication: Pre-crisis communication focuses on an organizational attempt to combine risk and crisis communication. From a CCC standpoint, once an assessment of community environment is conducted, any likelihood of danger or of impending disaster should be conveyed to community members, and especially to the community members who are most likely to be affected. For instance, the danger of illness outbreak, hurricanes, chemical spills, and water contaminants, and other dangers and disasters need to be communicated. The goal or hallmark of the pre-crisis communication is to help community members prepare, minimize, and/or avert crises altogether.

In the days following Hurricane Katrina, federal agencies focused on helping community members' preparedness by directing local emergency managers to encourage individuals to have emergency kits (FEMA, 2009; Seeger & Reynolds, 2007; Veil, Mitcham, & Sellnow, 2012). The kits are planned to help individuals and their families to withstand disasters before emergency help arrives or can be deployed. The emergency kits can include basic necessities, such as drinking water, food, medicine, and flashlights. Furthermore, risk and crisis communication are not mutually exclusive propositions in the sense that risk communication focuses on issues before a crisis and influences stakeholders' and publics' perception and expectations along with reaction to crisis communication in a given crisis situation (Veil et al., 2012). Risk communication is aimed at helping and persuading diverse publics to make certain behavioral changes. However, it must be understood that whether and how members of the diverse publics make the necessary changes may be affected by other factors. For example, risk communication in a health context has shown that compliance depends on the degree to which messages align with people's values, needs, and cultural beliefs (Institute of Medicine, 2002; Veil et al., 2012). In other words, simply talking about the risk of a potential crisis does not guarantee compliance. In fact, it has been suggested that developing or coming up with an appropriate message is less of a challenge than getting people to take appropriate or recommended actions suggested by the message. Thus, persuasion and gaining compliance through risk communication are challenging tasks (Veil, Littlefield, & Rowan, 2009; Veil et al., 2012).

Communication during a crisis: Crisis communication as explored in the previous section, focuses not only on pre-crisis but across all the three stages of crises: pre, during, and post. Most scholars have studied or written about crisis communication as organizations attempt to inform the public or stakeholders about crisis events and what such organizations are doing to address a particular crisis. However, from the AMCM perspective, and in particular for institutions like the CCC, crisis communication needs not only address the occurring crisis but should offer an opportunity for crisis managers to identify or to share with their stakeholders information about the crisis plan (i.e., pre-crisis). This approach allows stakeholders to determine for themselves whether an organization has taken reasonable steps to prevent a crisis from happening, one which consequently will impact how an organization is perceived as it handles a crisis.

However, communication during a crisis ought to focus on implementation of the crisis plan put in place prior to the crisis. Hence, from a CCC standpoint, crisis communication will focus on both risk communication and deployment of emergency workers and assistance to help community members restore some sense of normalcy in a chaotic environment. When a crisis occurs, there is an immediate demand for information to the extent that crisis communication ought to focus on helping the affected public reduce their level of uncertainties, for example, by providing the audience with an explanation of what happened so that they can take appropriate actions (Gilpin & Murphy, 2008; Lukaszewski, 2013; Reynolds & Seeger, 2005).

Crisis communication also extends beyond providing the audience with information about what happened or helping affected publics to cope physically—it should also inform how to cope emotionally or psychologically with a given crisis (Coombs, 2014). While most of the available literature has focused on crisis communication as a way to restore organizational image (e.g., Veil et al., 2012), CCCs' crisis communication must focus less on this (because they have no shareholders and the organization is less driven by profit motive) and more on safeguarding the community. That is, the CCC needs to act or to communicate with its public in a way that helps reduce dangers and harm from a given crisis while helping to restore the community to a situation where members can move beyond the crisis as quickly as possible. In other words, CCC crisis communication and management will be judged on the basis of how well the center is able to provide adequate information and to deploy available resources in a manner perceived as fair to all members of its community. This emphasis would place a tremendous burden on CCC's communicators in addressing the complex needs, cultures, and backgrounds of multiple stakeholders (i.e., community members). This approach creates a shift from the often taken-for-granted situation where authorities are believed to be in control of the communication process when they send messages and information through mass media without accounting for unique needs of the different audiences (Bridgeman, 2008; Gilpin & Murphy, 2008; Zaremba, 2015). At the same time, the one-way communication from mass media does not account for the audience feedback needed in the communication process. However, though the use of the traditional mass media may have been the norm in the past, today's average person or audiences are more sophisticated in how they receive and consume messages. More importantly, they want to actively engage in the communication process through the use of new and social media (Gruber, Smerek, Thomas, Hunt, & James, 2015; Jost, Banaji, & Nosek, 2004). Bridgeman (2008) suggests that the days of waiting till the end of the month or quarter before an organization takes action is over, and that people charged with crisis management, or public relations officers, need to monitor in real time new communication channels, such as blogs and international news. Perhaps one of the worst mistakes to make in communication with stakeholders is to assume that crisis managers must maintain strict control over information and access to such information. In other words, with varieties of new media, there is a greater need for crisis managers and, in particular, CCC leadership, to maintain transparency with the stakeholders, which will further enhance trust and trust building.

Audiences today want information right away, and they also want to exercise control over how information is shared within their own networks. All of this further complicates stakeholders' crisis communication and must be taken into consideration by communicators during the crisis process. Veil, Reynolds, Sellnow, and Seeger (2008) indicate that audience diversity in a crisis situation results or creates *uneven vulnerabilities*. At the same time, it should be emphasized that just because information is sent to stakeholders, that does not mean it is received or understood.

Post-crisis communication. Post-crisis communication represents the opportunity to inform the public about steps in place to prevent reoccurrence of a particular crisis. It also offers the opportunity for a given organization to assess what went well and what needs improvement. In retrospect, post-crisis communication provides a way to assess the effectiveness of risk and crisis communication and can include further risk communication to warn the public of any lingering dangers that may still lie ahead of a given crisis situation. Research in crisis management has never discussed post-crisis communication as an opportunity to celebrate accomplishments, perhaps because the crisis situation often evokes negative feelings from the audience. However, for organizations such as the CCC, it is important to celebrate with community members in terms of how collaboration between them and emergency workers and community volunteers helps curb the potential damage or destruction a given crisis might bring without such collaboration. This is important for two reasons. It acknowledges the role of stakeholders' or community members' contribution to crisis management and consequently helps them feel better about participating through volunteering in future crisis situations. It also helps to foster a sense of trust between the CCC and community members. The idea of celebrating accomplishments after a crisis is similar to the notion of *renewal*, suggested by Ulmer, Seeger, and Sellnow (2007) as an important aspect of post-crisis management beyond the focus on mere image restoration. The concept of renewal is not so much about accusation and blame, but instead about *instinctive, provisional, often value-based* *responses* to a crisis and an organization's ability to demonstrate good leadership by focusing on opportunities in a crisis, and to attempt to set in motion a vision for overcoming any crisis obstacle (Ulmer, Seeger, & Sellnow, 2007). Furthermore, communication in a post-crisis situation should be done in a way that sets in motion preparation for future crises and disasters. Hence, the AMCM approach is always an ongoing, continuous process because as one crisis ends, the anticipation of and planning for another, begins.

Other Issues/Factors Affecting the CCC and Stakeholders' Interactions

When dealing with multiple stakeholders, one needs to understand the uniqueness of each stakeholder group. Earlier, we stressed that the groups can be culturally diverse. In most communities there are ethnicity, religion, and language differences, which may also lead to cultural values and beliefs. For example, in our own community, there are three distinct ethnic minority groups: Asians, African Americans, and Hispanics/Latinos. However, when dealing with different groups, there is a tendency for mistrust and for perception of inequality, real or not. The level of mistrust influences how different stakeholders perceive or see themselves as part of the community. In an attempt to overcome this challenge, it is important that the CCC engage or select multiple spokespersons who are respected opinion leaders within their respective groups. These spokespersons, however, must be able to present clearly a consistent message using appropriate media. Although this approach appears to diverge from the established tradition in public relations and crisis management that insisted on having only one spokesperson, the critical element is *consistency* and not whether a message is presented by one or multiple persons. At the same time, consistency has been shown to allow an organization to be seen as legitimate and credible (Gilpin & Murphy, 2008; Massey, 2001).

Furthermore, beyond the physical distinguishable characteristics among multiple stakeholders, the notion of a vulnerable population has been introduced (see FEMA, 2004, as cited in Veil et al., 2012). A vulnerable population includes people with physical, medical, sensory, or other cognitive impairments. The notion of vulnerable populations has also been extended to the elderly, minority groups, non-English speakers, and people without transportation, along with members of lower socioeconomic classes. In particular, when disaster strikes, people in lower socioeconomic classes tend to be at a disadvantage because they have few options. For instance, they tend to be underinsured or have no insurance; they have less physically stable housing (Veil et al., 2012; Zakour & Harrell, 2003). Veil and colleagues (2012) also point to the dilemma of low socioeconomic status groups and their inability to afford emergency kits; hence they are limited in their preparedness for a crisis or disaster, which may also hinder their access to technology and hence risk communication messages disseminated via those technologies. At the same time, today's digital age also offers social media and other peer-to-peer technology platforms, such as Facebook and Twitter, that some groups prefer as a quicker and more reliable mode of information dissemination and acquisition, especially in crisis situations (Bridgeman, 2008). Chapter 3 provides a more in-depth discussion of technology use in crises.

No crisis communication will be effective without trust. However, minority individuals who are culturally diverse or view themselves as marginalized, and vulnerable populations, tend to be skeptical and do not necessarily trust information from warning messengers, and they also have the tendency to interpret information differently because of different cultural backgrounds. Furthermore, these groups of individuals have different perceptions of risk and use different channels of communication (Lachlan, Burke, Spence, & Griffin, 2009). Minorities also seek out information using different methods, as some rely on relatives and friends to make sense of risk messages. In particular, research notes that racial and ethnic minorities are not likely to accept risk communication from government officials without confirmation of the message within their personal networks (Lindell & Perry, 2004). For instance, part of the discussion in the aftermath of Hurricane Katrina was the possibility that perhaps there were not enough credible outlets to convey the threat severity of the hurricane to African Americans (Lachlan, Burke, Spence, & Griffin, 2009).

Therefore, to foster trust and to communicate emergency messages with not only multiple stakeholders but also with a culturally diverse audience, it is important to first acknowledge that differences exist and then proceed to take appropriate action to foster the level of trust. Having a good understanding of the level of differences in a multicultural environment can produce a good impetus for message design, for media selection, and for other interpersonal communication channels best suited to reach the underrepresented groups (Sellnow & Vidoloff, 2009; Veil et al., 2012; Williams & Olaniran, 1998). As we alluded to earlier, it needs to be stressed that the relationship develop trust with members of the disadvantaged or culturally diverse populace, and it needs to exist before a crisis or an emergency and should continue even after the crisis is over (i.e., ongoing). Waiting for an emergency or a crisis to strike before efforts to include underrepresented groups is a sure way to ruin crisis management efforts and any associated communication processes.

Language Differences

When there is a language difference within a community, it is paramount that a CCC or any crisis management program address this need. The attitude that English is the official language in the United States may not have much traction when it comes to good and ethical conduct during a crisis. It has been suggested that as high as 20% of people in the United States are non-English speakers (Shin & Bruno, 2003). In addition, Veil and colleagues (2012) presented U.S. Census Bureau data indicating that almost 47 million people in the United States do not speak English as their first language. Language differences will require crisis managers and practitioners and, in particular, the CCC to prepare ahead of emergencies and crises to recognize and put in place appropriate actions. For example, handouts and pamphlets may have to be presented in multiple languages to reach the desired audience. Similarly, providing public announcements and crisis messages on ethnic radio and on TV stations may also help. After all, some of the intended population may not be able to read. This is why multiple communication channels and messengers may help create the necessary communication redundancies needed to get risk-communication messages to the right people and in a way that they understand and can relate to (Olaniran & Williams, 2012). Furthermore, addressing language difference needs by a CCC can help reduce the complexity that language differences create and also help develop the necessary trust needed to ensure risk and crisis message compliance, especially when non-English-speaking ethnic minorities have problems trusting messages coming from an outsider. In essence, building and developing relationships with different stakeholders prior to crisis occurrence can help develop the necessary trust that is needed in facilitating effective communication. It is incumbent on CCC representatives to operate in a manner that helps different stakeholders view them in a positive light. This is in line with literature suggesting that crises create uncertain environments, and in times of uncertainty it is essential that an organization and its decision makers operate in a manner that allows stakeholders to see them as responsive to their respective needs—the organizational history of acting in a socially responsible manner goes a long way in this respect (Bridgeman, 2008; Padgett, Cheng, & Parekh, 2013). After all, when an organization neglects a certain stakeholder group, that group can create problems that will have a lasting negative impact on the organization (Freeman, 1984; Freeman & Gilbert, 1987; Gilpin & Murphy, 2008; Hearit, 2006; Zaremba, 2015).

Conclusion

Stakeholders are valuable to crisis management. They are not just bystanders in times of crisis. They have vested interests in the crisis, and, more importantly, they want to be involved in crisis management process. Thus, it is important that the CCC, as an organization, finds ways to partner with these groups throughout crisis management stages. When possible, we recommend engaging them to determine that their needs and expectations will pay good dividends to crisis practitioners as they deal with tasks of managing crises and disasters. Similarly, the CCC needs to recognize the importance of representing multiple stakeholders in its communications and in the general decision-making process. It should also realize that different stakeholders may utilize or respond to different communication media, and especially to the new media.

References

Avery, E. J., & Lariscy, R. W. (2010). FEMA and the rhetoric of redemption: New directions in crisis communication models for government agencies. In W. T. Coombs & S. Holladay (Eds.), *The handbook of crisis communication* (pp. 319–34). Malden, MA: Wiley-Blackwell.

Bland, M. (1998). *Communicating out of a crisis*. London: Macmillan Business.

Buber, M. (2002). From "Dialogue" (1932). In A. D. Biemann (Ed.), *The Martin Buber reader* (pp. 189–202). New York: Palgrave.

Bridgeman, R. (2008). Crisis communication and the net. *Crisis communication: practical PR strategies for reputation management and company survival* (pp. 169–177). Philadelphia: Kogan Page.

Coombs, W. T. (2014). *Ongoing crisis communication: Planning, managing, and responding* (2nd ed.). Thousand Oaks, CA: Sage.

Coombs, W. T., & Holladay, S. J. (2008). Comparing apology to equivalent crisis responses strategies: Clarifying apology's role and value in crisis communication. *Public Relations Review*, *34*, 252–57.

Coombs, W. T. (1993). Philosophical underpinnings: Ramifications of a pluralist paradigm. *Public Relations Review*, *19*(2), 111–119.

de Bussy, N. M. (2010). Dialogue as a basis for stakeholder engagement: Defining and measuring the core competencies. In R. Heath (Ed.), *Handbook of Public Relations* (pp. 127–44). Thousand Oaks, CA: Sage.

Federal Emergency Management Agency. (2004). *Are you ready? An in-depth guide to citizen preparedness*. http://www.fema.gov/areyouready/

Federal Emergency Management Agency. (2009). Post Katrina: What it takes to cut the bureaucracy and assure a more rapid response after a catastrophic disaster. Statement of Craig Fugate before the House Committee on Transportation and Infrastructure, Subcommittee of Economic Development, Public Buildings and Emergency Management U.S. House of Representatives. http://www.fema.gov/pdf/about/testimony/072709_fugate.pdf

Freeman, R. E. 1984. *Strategic management: A stakeholder approach*. Boston: Pitman.

Freeman, R. E., & Gilbert, D. R. 1987. Managing stakeholder relationships. In S. P. Sethi and C. M. Falbe (Eds.), *Business and society: Dimensions of conflict and cooperation* (pp. 397–423). Lexington, MA: Lexington Books.

Frooman, J. (1999). Stakeholder influence strategies. *Academy of Management Review*, *24*(2), 191–205.

Gilpin, D. R., & Murphy, P. J. (2008). *Crisis management in a complex world*. New York: Oxford University Press.

Gruber, D. A., Smerek, R. E., Thomas-Hunt, M. C., & James, E. H. (2015). The real-time power of Twitter: Crisis management and leadership in an age of social media. *Business Horizons*, 58(2), 163–172.

Hearit, K. M. (2006). *Crisis management by apology: Corporate response to allegations of wrongdoing*. New York: Routledge.

Heath, R. L. (1997). *Strategic issues management: Organizations and public policy challenges*. Thousand Oaks, CA: Sage.

Institute of Medicine. (2002). *Speaking of health: Assessing health communication strategies for diverse populations*. Washington, DC: National Academy Press.

Isaacs, W. (1999). *Dialogue and the art of thinking together*. New York: Doubleday.

Jost, J. T., Banaji, M. R., & Nosek, B.A. (2004). A decade of system justification theory: Accumulated evidence of conscious and unconscious bolstering of the status quo. *Political Psychology*, *25*(6), 881–919.

Lachlan, K. A., Burke, J., Spence, P. R., & Griffin, D. (2009). Risk perceptions, race, and Hurricane Katrina. *The Howard Journal of Communications*, 20, 295–309.

Leitch, S., & Motion, J. (2010). Publics and public relations: Effecting change. In *The Sage handbook of public relations*, 99–110. Thousand Oaks, CA: Sage

Lerbinger, O. (1997). *The crisis manager facing risk and responsibility*. Mahwah, NJ: Lawrence Erlbaum.

Lerbinger, O. (2012). *The crisis manager* (2nd ed.). New York: Routledge.

Lindell, M. K., & Perry, R. W. (2004). *Communicating environmental risk in multiethnic communities*. Thousand Oaks, CA: Sage.

Lukaszewski, J. E. (2013). *On crisis communication: What your CEO needs to know about reputation, risk and crisis management*. Brookfield, CT: Rothstein.

Massey, J. E. (2001). Managing organizational legitimacy: Communication strategies for organizations in crisis. *Journal of Business Communication*, 38, 153–83.

McDonald, L. M., Sparks, B., & Glendon, A. I. (2010). Stakeholder reactions to company crisis communication and causes. *Public Relations Review*, 36, 263–71.

McVea, J. F., & Freeman, R. E. (2005). A Names-and-faces approach to stakeholder management. *Journal of Management Inquiry*, *14*(1), 57–69.

Meserve, J. (2007, November 7). FEMA press secretary submits resignation, official says. http://www.cnn.com/2007/US/11/07/fema.official.resigns

Millar, F. E., & Beck, D. B. (2004). Metaphors of crisis. In D. P. Millar & R. L. Heath (Eds.), *Responding to crisis: A rhetorical approach to crisis communication* (pp. 153–166). Mahwah, NJ: Lawrence Erlbaum Associates.

Mitchell, R. K., Agle, B. R., & Wood, D. J. (1997). Toward a theory of stakeholder identification and salience: Defining the principle of who and what really counts. *Academy of Management Review*, *22*(4), 853–86.

Mitroff, I. I. (2004). *Crisis leadership: Planning for the unthinkable*. Hoboken, NJ: John Wiley.

Olaniran, B. A., Scholl, J. C., Williams, D. E., & Boyer, L. (2012). Johnson and Johnson phantom recall: A fall from grace or a re-visit of the ghost of the past. *Public Relations Review*, *38*, 153–55.

Olaniran, B. A., & Williams, D. E. (2001). Anticipatory model of crisis management: A vigilant response to technological crises. In R. L. Heath (Ed.), *Handbook of public relations* (pp. 487–500). Thousand Oaks, CA: Sage.

Olaniran, B. A., & Williams, D. E. (2012). The need for an Anticipatory Perspective in Crisis Communication. In. B. Olaniran, D. Williams, & T. Coombs (Eds.), *Pre-crisis planning, communication, and management: Preparing for the inevitable* (pp. 1–16). New York: Peter Lang

Padgett, D. R. G., Cheng, S. S., & Parekh, V. (2013). The quest for transparency and accountability: Communicating responsibly to stakeholders in crises. *Asian Social Science*, 9(9), 31–44.

Pearson, C. M., & Mitroff, I. I. (1993). From crisis prone to crisis prepared: A framework for crisis management. *Academy of Management Executive*, *7*, 48–59.

Pfeffer, J., & Salancik, G. R. (1978). *The external control of organizations: A resource dependence perspective*. New York: Harper & Row.

Phillips, K., Chernoff, A., Oppenheim, K., & Jeras, J. (2007, October 29). Fire deaths, damage come into focus as evacuees cope. http://www.cnn.com/2007/US/10/26/fire.wildfire.ca/

Rawlings, B. L., & Bowen, S. A. (2005). Publics. In R. Heath (Ed.), *Encyclopedia of public relations* (pp. 718–21). Thousand Oaks, CA: Sage.

Ray, S. J. (1999). *Strategic communication in crisis management lessons from the airline industry*. Westport, CT: Quorum.

Reynolds, B., & Seeger, M. W. (2005). Crisis and emergency risk communication as an integrative model. *Journal of Health Communication*, *10*, 43–55.

Scholl, J., Williams, D., & Olaniran, B (2005). Preparing for terrorism: A rationale for crisis communication center. In H. O'Hair, R. Heath, & G. Ledlow (Eds.), *Community preparedness and response to terrorism* (pp. 243–68).Westport, CT: Praeger.

Seeger, M. W., & Reynolds, B. (2007). Crisis communication and the public health: Integrated approaches and new imperatives. In M. W. Seeger, T. L. Sellnow, & R. R. Ulmer (Eds.), *Crisis communication and the public health* (pp. 3–20). Cresskill, NJ: Hampton.

Sellnow, T. L., & Vidoloff, K. G. (2009, September). Getting communication right. *Food Technology*, 63(9), 40–45.

Senge, P. M. (2006). *The fifth discipline* (2nd ed.). London: Random House.

Shin, H., & Bruno, R. (2003). *U.S. Census Bureau language use and English-speaking ability: 2000*. http://www.census.gov/prod/2003pubs/c2kbr-29.pdf

Stephens, K. K., Barrett, A. K., & Mahometa, M. J. (2013). Organizational communication in emergencies: Using multiple channels and sources to combat noise and capture attention. *Human Communication Research*, 39(2), 230–251.

Stephens, K. K., Malone, P. C., & Bailey, C. M. (2005). Communicating with stakeholders During a Crisis Evaluating Message Strategies. *Journal of Business Communication*, 42(4), 390–419.

Tazin, S. N. M., & Kaur, K. (2014). Public relations challenges in crisis management: Discovering strategic crisis communication and stakeholder reactions in a Malaysian government-linked company. *Media Asia*, *41*(1), 25–30.

Ulmer, R. R., & Sellnow, T. L. (2000). Consistent questions of ambiguity in organizational crisis communication: Jack in the Box as a case study. *Journal of Business Ethics*, 25, 143–55.

Ulmer, R. R., Seeger, M. W., & Sellnow, T. L. (2007). Post-crisis communication and renewal: Expanding the parameters of post-crisis discourse. *Public Relations Review*, 33(2), 130–34.

Veil, S. R., Littlefield, R. S., & Rowan, K. E. (2009). Dissemination as success: Local emergency management communication practices. *Public Relations Review*, 35(4), 449–451.

Veil, S. R., Mitcham, C., & Sellnow, T. (2012): Lost in translation: how emergency managers communicate preparedness with vulnerable populations. In. B. Olaniran, D. Williams, & T. Coombs (Eds.), *Pre-crisis planning, communication, and management: Preparing for the inevitable* (pp. 201–225). New York: Peter Lang

Veil, S. R., Reynolds, B., Sellnow, T. L., & Seeger, M. W. (2008). Crisis & emergency risk communication as a theoretical framework for research and practice. *Health Promotion Practice*, 9(4), 26S–34S.

Williams, D., & Olaniran, B. A. (1998). Expanding the crisis planning function: Introducing elements of risk communication. *Public Relations Review*, 24(3), 387–400.

Wong, I., & Hung, Chun-Ju, F. (2008). Organization-stakeholder relationships, crisis responsibilities, and crisis response strategies. In B. Van ruler, A. T. Vercic, & D. Vercic (Eds.), *Public relations metrics: Research and evaluation* (pp. 207–34). London: Routledge.

Zakour, M., & Harrell, E. (2003). Access to disaster services: Social work interventions for vulnerable populations. *Journal of Social Service Research*, 30(2), 27–54.

Zaremba, A. J. (2015). *Crisis communication: Theory and practice*. New York: Routledge.

Exhibit on multiple stakeholders

(http://www.fbi.gov/boston/press-releases/2013/statement-by-special-agent-in-charge-richard-deslauriers-regarding-information-sharing)

Updates

November 21, 2013

The *Cyclospora* illness outbreaks during the summer of 2013 have ended, based on traceback and epidemiologic investigations conducted by the U.S. Food and Drug Administration (FDA) and the Centers for Disease Control and Prevention (CDC) as well as state and local public health authorities in Iowa, Nebraska, Texas, and other states. According to the CDC, reported cyclosporiasis cases have returned to baseline levels.

Epidemiologic and traceback investigations by the CDC, the FDA, and the states of Iowa and Nebraska, had linked some illness clusters in the outbreaks in those states to salad mix supplied by Taylor Farms de Mexico, S. de R.L. de C.V., to Olive Garden and Red Lobster restaurants, both owned by Darden Restaurants.

The FDA, in conjunction with the CDC, the Mexican government, and Taylor Farms de Mexico, conducted an environmental assessment at Taylor Farms de Mexico facilities in Guanajuato, Mexico, from August 12 to 19, 2013. The team found that conditions and practices observed at these facilities at the time of the assessment were in accordance with known food safety protocols.

The Food and Drug Administration along with the Centers for Disease Control and Prevention and state and local officials are investigating a multistate outbreak of *Cyclospora* infections. We are moving quickly to learn as much as possible and prevent additional people from becoming ill. We recognize that people will be concerned about this outbreak, and we will continue to provide updates and advice.

What Is the Problem and What Is Being Done About It?

The *Cyclospora* illness outbreaks during the summer of 2013, have ended, based on traceback and epidemiologic investigations conducted by the FDA and the CDC as well as by state and local public health authorities in Iowa, Nebraska, Texas, and other states. According to the CDC, reported cyclosporiasis cases have returned to baseline levels.

As of October 29, 2013, CDC had been notified of 631 cases of *Cyclospora* infection from 25 states: Arkansas, California, Connecticut, Florida, Georgia, Illinois, Iowa, Kansas, Louisiana, Massachusetts, Michigan, Minnesota,

Missouri, Nebraska, New Hampshire, New Jersey, New York (including New York City), Ohio, Pennsylvania, South Dakota, Tennessee, Texas, Virginia, Wisconsin, and Wyoming.

On July 30, 2013, the states of Iowa and Nebraska announced that their analysis indicated that the outbreak in those states was linked to a salad mix.

The FDA traceback investigation has confirmed that the salad mix identified by Iowa and Nebraska as being linked to the outbreak of cyclosporiasis in those states, was supplied to restaurants in those states by Taylor Farms de Mexico, S. de R.L. de C.V., a processor of foodservice salads. The FDA traceback investigation found that illness clusters at restaurants were traced to a common supplier, Taylor Farms de Mexico, S. de R.L. de C.V. The restaurants in Iowa and Nebraska include Olive Garden and Red Lobster, both of which are owned by Darden Restaurants. The FDA's investigation has not implicated consumer packages sold in grocery stores.

The Iowa Department of Inspections and Appeals (DIA) and the Nebraska Department of Health and Human Services have announced that they believe the contaminated salad is no longer in the food supply in those states. The last date that someone who had eaten in one of these restaurants in those states reportedly became ill with cyclosporiasis was on July 2. The typical shelf life for a salad mix is up to 14 days.

As a result of this investigation, the FDA increased its surveillance efforts on green leafy products exported to the United States from Mexico.

On August 12, 2013, Taylor Farms de Mexico officially informed the FDA that, as of August 9, 2013, the company had voluntarily suspended production and shipment of any salad mix, leafy green, or salad mix components from its operations in Mexico to the United States. The firm committed not to resume production and shipping of these products from its operations in Mexico without the FDA's concurrence. Only the salad mix was been implicated in the outbreak of cyclosporiasis in Iowa and Nebraska. This voluntary action went beyond the implicated salad mix and includes iceberg lettuce, romaine lettuce, green leaf lettuce, red cabbage, green cabbage, and carrots. The action of Taylor Farms de Mexico exemplified the company's cooperation with federal and state officials throughout this complicated investigation.

The Mexican government's food regulatory authorities, the Federal Commission for Protection against Sanitary Risks (COFEPRIS), the National Agro-Alimentary Health, Safety and Quality Service (SENASICA), and state-level agricultural authorities also collaborated with the FDA in the investigation of this outbreak.

From August 12 to 19, 2013, the FDA, with the cooperation of the Mexican government authorities and Taylor Farms de Mexico, S. de R.L. de C.V., conducted a thorough environmental assessment at Taylor Farms de Mexico's processing facility and five farms identified through the *Cyclospora* outbreak's traceback investigation. The team found that conditions and practices observed at these facilities at the time of the assessment were in accordance with known food safety protocols. A previous inspection, in 2011, of the processing facility of Taylor Farms de Mexico, S. de R.L. de C.V. conducted by FDA found no notable issues.

As a result of the environmental assessment and the FDA's thorough review of a product sampling plan for *Cyclospora* put in place by Taylor Farms de Mexico, S. de R.L. de C.V., the FDA agreed to the firm resuming operations. The firm has committed to a comprehensive *Cyclospora* sampling program for leafy green and other products from their farms and processing facility in Mexico. This will include both sampling of their products and water and continued monitoring of the sanitary conditions of their facilities.

On August 25, 2013, Taylor Farms de Mexico, S. de R.L. de C.V., resumed production and shipment of salad mix, leafy greens, and salad mix components from its operations in Mexico to the United States.

Epidemiologic and traceback investigations by state and local public health authorities in Texas as well as the FDA and the CDC indicate that some illnesses among Texas residents were linked to fresh cilantro grown in Puebla, Mexico. Investigations conducted by the CDC at three unrelated restaurants and one grocery store in Texas found that fresh cilantro was the only ingredient that was consumed by all persons in all of the four locations during the 7–14 day incubation period for the parasite.

A CDC epidemiologic study among persons who reported eating at a single Mexican-style restaurant in Fort Bend County, Texas, compared foods eaten at the restaurant by 25 ill persons and 65 healthy persons. Ingredient-level analyses were conducted using meal consumption data and recipes. Four fresh produce ingredients were significantly associated with illness: cilantro, whole onions, garlic, and tomatoes. Since these ingredients were served together in many different dishes, such as salsa, the statistical analysis could not definitively identify which of these ingredients was independently associated with illness. However, fresh cilantro was the only ingredient consumed by all 25 ill persons in this study. Consumption of lettuce was not associated with illness in this analysis, and traceback information indicated that lettuce served in this restaurant was not from Taylor Farms de Mexico.

Preliminary traceback information indicated that fresh cilantro harvested from the Puebla, Mexico area, was supplied to restaurants and the grocery store in Texas during the period when the persons who became ill purchased it. The FDA is working with Mexican authorities to determine if there is a common farm or growing area in Puebla, Mexico, which could have provided fresh cilantro to the three restaurants and the grocery store associated with the clusters of ill persons in Texas.

As a result of this investigation, the FDA has increased its surveillance efforts on cilantro products exported to the United States from Mexico. The FDA will continue to work with its federal, state, and local partners in the investigation.

What Is *Cyclospora*?

Cyclospora cayetanensis is a parasite composed of one cell, too small to be seen without a microscope. This parasite causes an intestinal infection called cyclosporiasis.

Cyclospora is acquired by people ingesting something—such as food or water—that was contaminated with the parasite. *Cyclospora* needs time (days to weeks) after being passed in a bowel movement to become infectious for another person. Therefore, it is unlikely that *Cyclospora* is passed directly from one person to another.

For more information on *Cyclospora*: http://www.cdc.gov/parasites/cyclosporiasis/ disclaimer icon

Who Is at Risk?

People can become infected with *Cyclospora* by consuming food or water contaminated with the parasite. People living or traveling in countries where cyclosporiasis is endemic, including certain tropical or subtropical regions of the world, may be at increased risk for infection.

What Are the Symptoms?

The time between becoming infected and becoming sick is usually about one week. *Cyclospora* infects the small intestine (bowel) and usually causes watery diarrhea, with frequent, sometimes explosive, bowel movements. Other common symptoms include loss of appetite, weight loss, stomach cramps/pain, bloating, increased gas, nausea, and fatigue. Vomiting, body aches, headache, fever, and other flu-like symptoms may be noted. Some people who are infected with *Cyclospora* do not have any symptoms. If not treated, the illness may last from a few days to a month or longer. Symptoms may seem to go away and then return one or more times (relapse). It's common to feel very tired.

What Do Consumers Need to Do?

Consumers should always practice safe food handling and preparation measures. Wash hands, utensils, and surfaces with hot, soapy water before and after handling food. Fresh produce should be thoroughly washed before it is eaten.

If more specific information becomes available, the FDA and the CDC will share it with the public, along with any steps consumers can take to prevent illnesses.

Who Should Be Contacted?

Contact your healthcare provider if you have diarrhea that lasts for more than three days.

The FDA encourages consumers with questions about food safety to call 1-888-SAFEFOOD or consult the fda.gov website: www.fda.gov.

The information in this release reflects the FDA's best efforts to communicate what it has learned from the manufacturer and the state and local public health agencies involved in the investigation. The agency will update this page as more information becomes available.

From FDA:

Additional Resources

- http://www.foodpoisonjournal.com/food-poisoning-watch/hepatitis-a-vaccines-cost-the-lubbock-health-department-over-250000/#.Ulgue-FCshcY
- http://www.wave3.com/story/15475307/umc-foots-bill-for-hep-a-vaccines?clienttype=printable
- http://www.foodsafetynews.com/2014/02/retailer-list-in-9-million-pound-beef-recall-expands/#.Uv6MSPldWrk
- http://www.reuters.com/article/2014/02/13/us-energy-crude-derailment-idUSBREA1C13120140213
- http://www.fda.gov/Food/RecallsOutbreaksEmergencies/Outbreaks/ucm361637.htm

· 3 ·

COMMUNITY RESOURCES FOR CCCS

Chapter Highlights

- What counts as resources in crises and in disasters
- The aim of resources in crisis and disaster readiness and management
- How resources can be coordinated by a CCC
- The role of opinion leaders as resources in crisis communication

Introduction

Now that the rationale for the CCC has been established, it is important to address how to make it work in a manner that would allow it to help the community address the pressing needs of planning, preparation, and crisis management. There is no point in having a CCC if it cannot function effectively. In an attempt to do this, this chapter focuses on identifying varieties of resources available within the community and on discussing how to mobilize them in the crisis process. Furthermore, each of the addressed resources is discussed within the context of the AMCM model and how they fit into the CCC. First, however, it is important to define or conceptualize what is meant by *resources*.

Resources

Because the goal of CCCs is to help the community prepare and to be ready for a crisis, it is necessary that the preparation takes into account the best way to coordinate and to utilize available resources at the community's disposal. The notion of human resources, in particular, can be traced back to the management and organization behavior fields, which focus on how people within organizations behave individually and in group settings within the confines of their respective organizations (e.g., structure, process, and culture) (Armstrong & Taylor, 2014). The focus of any resource-based theory, such as human resources management, is on achieving a desired end. Resources thus include people, personnel, or human capital along with equipment (e.g., assets) and even the nature of an organization, including *size*, *age*, *technology*, and *environment—internal and external*—in an attempt to impact performance (Armstrong & Taylor, 2014). In essence, for most organizations, resources include anything that can be used to gain and to sustain a competitive advantage (e.g., Armstrong & Taylor, 2014; Barney, 2012). Thus, the organizational context (internal and external environment) is central to the AMCM. Even though a CCC is not a traditional organization focusing on gaining financial or competitive advantages; its goal is similar to that of traditional organizations in the sense that it needs to find, acquire, and then mobilize its resources to plan, prepare, and manage both crises and disasters. Therefore, through the help of a CCC as a central location for crisis prevention and management, it is imperative that the center evaluate resources in a way to optimize its effectiveness and relevance within the community. In addition to conducting a thorough analysis of all of the factors related to specific communities about crisis potential, a way to deploy available resources to overcome a crisis when it occurs must be put in place. Furthermore, the AMCM encourages crisis experts and practitioners not only to develop and engage in systematic review of all crisis concerns, but also to provide appropriate and effective response strategies to confront crises. As a result, it is believed that the greater the crisis response, the greater control crisis experts will have in addressing a particular crisis (Olaniran & Williams, 2001, 2004; Scholl, Williams, & Olaniran, 2005).

CCC staff or people

Earlier, we addressed the staff composition of a CCC. CCC team composition is important because it is the team's job to anticipate potential crises and to put in place the mechanisms to prevent them or help in mitigating their impact once they become a reality. Therefore, it is paramount that once a community's predisposition to certain crises is determined, there are individuals in the CCC who are knowledgeable in how to address them. It is prudent for a CCC to consist of medical experts, scientific officers, technologically savvy individuals, safety experts, among others, who are able to act as liaisons between a CCC and the community (see also, Seeger, Sellnow, & Ulmer, 2003; Zaremba, 2010). To this end, it is imperative that the team of experts staffing a CCC is able to brainstorm or to identify a community's exposure to different types of crises and disasters in an attempt to safeguard or to prepare and manage them (Jaques, 2007; Lukaszewski, 2013; Zaremba, 2010). In addition to staff and CCC personnel, the role of stakeholders will also be addressed later in the chapter.

Emergency response team

Community resources are crucial for the effective management of any crisis. However, the emergency response teams are at the center of any crisis response team and their plans. In any crisis, the first several hours are crucial to crisis management. Specifically, FEMA indicates that for the first 72 hours following a crisis, one must take into account how communities and their residents would survive in the absence of major utilities, such as electricity, gas, water, and phones, which are likely to be inaccessible at the onset of the crisis and may not be immediately restorable. Green and Teper (2006) indicate that "reliance upon community resources may be the most valid way of providing protection against the threat of disaster" (p. 50). During the onset of a crisis, the responsibility to reduce fatalities and to care for the injured (i.e., health and well-being) traditionally falls on the shoulders of the respective local communities; to the extent that it is prudent, the CCC plans or puts in place a program that helps mobilize resources, integrating nonprofit organizations, such as churches, libraries, food pantries, or shelters, into a crisis plan (FEMA, 2001; Gomez & Turoff, 2007; Veil, Mitcham, & Sellnow, 2012). This is important, as situations such as hurricanes Sandy, Katrina, and Isaac suggest that external help may be late in arriving, and, consequently, communities are charged with the responsibility of dealing with crises and disasters alone until help arrives. This further reinforces the need to have a crisis plan in place before a crisis occurs. From the AMCM standpoint, CCC and respective crisis plans should be unique to the type of crisis and different crises may require different levels of expertise or knowledge. In essence, this chapter identifies a few community resources that can be mobilized or deployed in a concerted and coordinated effort with CCCs to deal with crises and with their respective roles. The ultimate goal is to provide a functional checklist or tools that can be adapted to each community's need and demands in their crisis plan.

Crisis Response Teams

Personnel

Community resources in terms of crisis response teams should consist of individuals who already serve the residents in a specialized and emergency role, such as public safety, public health organizations, and nonprofit organizations, and who respond when a crisis occurs and until external help arrives (Gomez & Turoff, 2007; Seeger, Sellnow, & Ulmer, 2003; Zaremba, 2010). These responders need to communicate and to coordinate activities with their team based on the circumstances surrounding the crisis. Community crisis response teams can also offer valuable services to external relief organizations because they know the environment and understand how and where to access particular information and resources (Basso, 1996). This is particularly important to help expedite the implementation of an action or crisis plan. Furthermore, because of the knowledge of their communities, local community crisis response teams serve another critical role, which involves the ability to identify vulnerable populations (i.e., children, the elderly, and people with disabilities) and the attempt to get to them right away (Veil, Mitcham, & Sellnow, 2012), or design appropriate messages to reach them. The CCC and its crisis response teams, through their volunteers, can also help in mobilizing other grassroots organizations that can assist citizens and affected populations in the preservation and maintenance of the health and well-being in times of crises.

Outside of the traditional emergency response teams, it is imperative that other individuals and institutions within the community are involved in the crisis response team. For example, libraries, shelters and food pantries, respected leaders in the community, and clergies and religious leaders, along with their place of worship, can offer valuable services to the complex requirement of the crisis response demands.

Library

The public library as a community resource can serve a valuable role during crisis planning and crisis management. The public library has the staff, resources, and expertise to address the need for training and can provide reliable and relevant information in any crisis situation. For instance, the librarians can facilitate community-wide training efforts in preparation for a crisis and in addressing the needs of the publics during a crisis (Will, 2001). The library can identify its own role in community recovery through information gathering, organization, and dissemination before, during, and after a crisis, according to Will (2001). However, librarians must not wait to be asked to become involved. The public library has invaluable access to and is a key repository for valuable information and databases, which allow quick access to national government protocol or crisis preparation guides, such as those put out by DHS and FEMA. With the aid of grants from such foundations as the Bill and Melinda Gates Foundation, most public libraries are equipped with Internet computers to use at will. A proactive public library can serve as an active partner in communities' emergency crisis plans and preparedness because it can collect valuable information and present it in a summarized, meaningful, and locally useful manner to its residents. For instance, after a disaster, there is usually very limited access to traditional communication channels because of power outages and lack of access to phones; thus, local public libraries, which are equipped with free access to computers and the Internet, can serve as the mode of communication and access to information in most communities soon after disasters (Bishop & Veil, 2013). The library can also become a go-to place for those charged with the city's responsibilities of planning and putting in place emergency plans (Will, 2001). Moreover, because most local communities, and especially those in rural communities, lack crisis plans or terrorism preparedness, assistance by public libraries would be welcomed by such local governments (Bishop & Veil, 2013; Green & Teper, 2006).

In terms of specifics, a public library should attempt to provide accurate data and to identify authoritative and useful Internet links for city and county planners in areas of interest (Will, 2001). Personnel training is mentioned frequently for emergency workers in most city plans. Libraries and their librarians have the facility and the skills to help make it a reality, whether by gathering and disseminating information or by helping identify appropriate experts to speak at such events. Results from a 2010–11 survey of public libraries suggested that about 92% of the respondents reported helping people to navigate government websites (Bertot, McDermott, Lincoln, Real, & Peterson, 2012). After all, government forms for disaster recovery and assistance programs are all online. The problem, however, is that some people do not have the technical literacy to use them; thus library officials are able to assist in this area as individuals attempt to get the necessary help (Veil & Bishop, 2014; Mandel, Bishop, McClure, Bertot, & Jaeger, 2010). The library may also serve as the place to conduct mock crisis exercises. For the general public, there is the need to know and to feel that they have the necessary information to protect themselves and their families from impending threats and dangers that a crisis entails, and the library is a good place to get the necessary information. Will (2001) argues that people will want information in terms of what to do and what not to do, emergency telephone numbers, and reliable sources for news updates, including school closings for their children, traffic information (e.g., detours and reroutes), business closings, where to seek shelters, and sources of food and water.

Perhaps one of the best ways a public library can help disseminate its crisis information is through a designed website, which must be ready to go live prior to a crisis. The site would need to be user-friendly, and valuable information should not be buried in a plethora of textual information. If bandwidth allows, there needs to be short audio and video clips. More importantly, the site must be designed with the ability to edit information remotely, especially if the library itself is located in close proximity to a crisis. Green and Teper (2006) point to the need to digitize information in case a library is destroyed during a disaster so that information may still be available to users. Will (2001) suggests that the site must have the capacity to support multiple users without crashing. One cannot overemphasize the importance of the site's availability to community members; otherwise panic may botch a good crisis plan. Hurricanes Katrina and Rita depicted commotion as people took things into their own hands and everybody tried to evacuate the cities all at once, consequently causing logjams on the highways and on other city infrastructures. We concur with Will (2001) that while the library can serve other roles in crisis management, its most valuable role is in gathering, compiling, and disseminating information, and no other center is as equipped as the library to perform such roles.

The roles of public library services in crisis and disaster situations also include serving as shelters and housing city command centers for emergency responders; distributing food and supplies, providing a place to recharge electronics and communication devices, and assisting with the completion of FEMA insurance claims (Bishop & Veil, 2013; Jaeger, Langa, McClure,

& Bertot, 2006) The increasing role of libraries as critical resources is indicated in another study, which found that although 35% of Louisiana public libraries were closed during Katrina, the overall number of visitors only went down by 1%, meaning individuals sought help from libraries not in their immediate community, especially when theirs was closed (Skinner, 2007). Furthermore, the critical role of public libraries as a primary resource for disaster response was emboldened on January 7, 2011, when FEMA changed its policy to allow libraries to be eligible for funding for temporary relocation during major disasters and emergencies (Bishop & Veil, 2013; Kelley, 2011).

It is important to note that the literature is rife with the increasingly important role of libraries in post-crisis and disaster management. However, very little is known or presented about the role that libraries can play in crisis and disaster planning. This is important. In staying with the tenet of the AMCM, libraries and their officers need to be involved in the crisis and disaster planning stage to identify services offered by local libraries to assist in crisis management, in putting crisis plans together, and in understanding areas where even libraries may need assistance. For instance, Green and Teper (2006) suggest that large libraries are able to afford backup generators, which most smaller libraries may not be able to afford. The question looming in such an instance for a CCC as it puts its planning and crisis preparedness together could be this: Is there any nearby library that can perform needed functions in such a scenario or does the community need to figure a way to purchase a generator in an attempt to address the needs of its people? Other pre-crisis and disaster-planning roles of the library would include practice drills to prepare to implement or to revise the crisis plan. Also, if a library is to serve as a temporary shelter, how would this effort be handled? Similarly, libraries need not only be prepared to handle short-term emergencies, such as securing patrons during tornado warnings, they also need to have long-term plans. The plan also needs to focus on training their employees in how to handle emergencies, secure the building, and keep themselves safe (Bishop & Veil, 2013; Green & Teper, 2006). As a matter of fact, the need for libraries to have long-term disaster plans was emphasized after the 2011 tornadoes by Alabama Public Library Service (Bishop & Veil, 2013).

Shelters

Shelters are places that provide a temporary stay for people in need. Shelters could offer a vital role in community crisis action plans. While most shelters provide some kind of emergency and temporary assistance and aid to people and displaced families or a community residence during a crisis, shelters can offer valuable assistance to crisis and disaster response teams during planning and preparation. They can offer community crisis planners information regarding the number of spaces and beds available, such that crisis planners would not wait till the time of crisis before gathering such information. This information should be readily available in a CCC database, such that when disaster strikes, the CCC and response teams would know exactly how many people to send where or the number of spaces available at different shelters. This would help expedite and smooth an otherwise logistically challenging or time-consuming function. Some renowned nonprofit organizations, such as the Salvation Army and the Red Cross, offer mobile units that go out and pick up individuals to bring to the shelter. For instance, some mobile units in times of extreme weather go out to perform this role and can easily be deployed in times of crises and disasters.

Food pantries and churches

Food pantries often offer the same or similar services as shelters, but their primary aim is to provide food for community members. The role of food pantries is especially important to crisis planning and management. Food pantries can provide food and toiletries for displaced members of the public in an attempt to help them to become self-sufficient or to provide for themselves and members of their families until a crisis blows over. At times, food pantries are places of last resort for community members, especially when there is a power outage that renders grocery stores useless in their ability to serve patrons. Food pantries often have in stock packaged or canned goods that are imperishable and with longer shelf life than fresh produce available at local grocery stores. Therefore, they can help the public tremendously in times of crises. Like shelters, information regarding the capacity or the number of people who can be sustained by any given food shelter needs to be readily available. In conforming to the anticipatory model, it would help to identify all food pantries within the community, and, if at all possible, the type of services they render should be published ahead of time and posted on the website available through the library and through the crisis response teams. A case in point is the Lubbock, Texas, community, which offers the Breedlove Food Dehydration Plant. Though not necessarily a food pantry per se, it provides food and soup packages for relief assistance around the globe. Such a plant can make its services available to a local community in times of crises. It may also work with other food pantries by delivering supplies directly to them. Furthermore, there are locally identifiable food pantries, but there are also those that may not be known or be readily accessible. Therefore, from an AMCM perspective, CCCs should have points of contacts with local food pantries along with their respective capacities in the database, such that when the need arises for these pantries, their services can be deployed relatively with ease. At the same time, knowing each food pantry's operational capacities can allow CCC staff to plan ahead regarding whether to negotiate with other communities to assist, if need be. In essence, all efforts must be made to identify different pantries and their services, and perhaps create a list or database that can be easily accessible to community members. For example, some local churches or places of worship offer food pantry–like services within the community. Some of them are known as Christ's Cupboard and cater to the homeless and economically disadvantaged and other vulnerable populations. The services rendered by churches in food supplies would add to those offered by traditional food pantries and allow for an increase in reach to members of the community and hence can become invaluable during a crisis.

Churches

In addition to its social support functions, such as providing a food pantry, churches also serve a fundamental function of providing critical information or counseling in crisis management (i.e., pre-, during, and post-). For instance, it has been suggested that people attempt to seek information from varieties of sources in times of crises (Palen & Liu, 2007). However, individuals, for the most part, tend to look to their own personal social networks—friends and family and other trusted sources, such as clergy—to determine and make sense of information and risk communication from formal or government sources, and then make decisions about their own response strategies, such as evacuation and compliance to those risk communication messages (Mileti, Sorensen, Sorensen, & Sutton, 2002; Wachtendorf, 2004; Weick, 1979, 1996). Previous research suggests that certain segments of the population, namely, ethnic minorities, have a distrust of *risks* or *warning messages* emanating from government officials; thus, they tend to lean on their personal networks for confirmation or the credibility of such messages (Lindell & Perry, 2004; Veil, Mitcham, & Sellnow, 2012). In other words, individuals are likely, and often collectively, to decide with their extended networks (e.g., family, friends, and neighbors) regarding their evacuation behavior based on how they interpret warnings, sources, and modes of communication (Drabek, Key, Erickson, & Crowe, 1975). The implications from this study are that individuals are more likely to trust people they believe to have their best interest at heart, thus church clergies are placed at the center of such a role. Hearit (2006) also argued that there are times during a crisis when individual confidentiality needs to be protected; there are no better places for that than a place of worship, where victims can feel safe and secure, knowing that whatever information they discuss would be held with strict confidentiality. Thus, a CCC needs to have as part of its resources a list of churches categorized by denominations to make sure all religions are covered as much as possible and that the information be easily available for community members. In other words, the CCC must anticipate how publics are likely to react within their networks rather than discarding these personal networks as irrelevant, because they are crucial to risk communication and message effectiveness. Furthermore, churches may be essential in sorting out dissonance or risk communication messages believed to be in conflict with religious beliefs and or obligations. Churches and faith-based organizations provide other important services, such as planned relief services like child day care, which allows parents and or crisis victims and survivors to move beyond crisis occurrence or to cope with crises. For instance, the child day care services offered by these organizations and groups often allow parents to attend to the completion of insurance claims forms and other necessary paperwork (Gomez & Turoff, 2007).

Hospitals and clinics

Hospitals and clinics occupy a central role in crisis management. One must refrain from trying to exaggerate the importance of one institution over the other in terms of crises, but there is no doubt that the responsibilities of crisis response teams and, in particular, of emergency and first responders include the ability to get community members and affected publics the right medical care. Hence, hospitals and clinics offer expertise in dealing with illnesses and injuries resulting from a particular crisis. Therefore, they should be included as part of a crisis response team and community resources. The EMS system needs to have a place to take patients or injured individuals to and they must possess full confidence that affected individuals will receive proper care. Under normal circumstances, hospitals know how to care for patients, but they need to be ready to handle the load of an increased number of patients in times of crises and disasters. Thus, hospitals and clinics need to have the necessary staff and the ability to triage publics affected by crises. They must train their staff in how to work under extreme pressure without complicating the medical needs and requirement of the patients. Hospitals will serve as the first place of care in cases of accidents, epidemics, pandemics, and other needs. In extreme cases, such as epidemics and pandemics, they must be able to link themselves with government establishments, such as the Centers for Disease Control (CDC), to communicate needs, assessment, and other protocols in the attempt to give the affected publics the best care possible while preventing the continued spread of disease within the community. If quarantine is needed, hospitals should be able to do it without any problems. This is why it is important that a respectable member of the medical community is staffed in the CCC to coordinate communication and delivery of care. More importantly, such an individual is needed to conduct need analysis and assessment to avert potential crises. Stewart (2013) discussed how two medical doctors in different states coordinated their efforts to help the victims of the Boston Marathon bombing. Specifically, Ron M. Walls of the Brigham and Women's Hospital in Boston, worked with Richard Zane, a former colleague, who dealt with the victims of the Aurora, Colorado, shootings to establish the new trauma threshold of handling 23 patients per hour. According to Walls, "We raised our game. We elevated our game because of what had happened in Colorado. Can you take care of 23 patients in a very short period of time, in the course of an hour? We were able to deploy multiple teams and have that readiness" (Stewart, 2013). In essence, the lesson learned from the Aurora shooting was instrumental in helping prepare the people of Boston for a crisis while also helping in the care delivery after the onset of the crisis (see http://www.thedenverchannel.com/news/movie-theater-shooting/aurora-response-is-serving-as-textbook for emergency-response).

Furthermore, in cases of rural communities where the best care is only available through small clinics, it is essential to figure out a way to work with a larger and nearby community with a full-service hospital that can deliver the needed services. The logistics of transportation of the affected populace to the necessary hospital would have to be addressed well ahead of a given crisis. Public health and safety are necessary for both proactive and reactive efforts, especially in crisis preparedness, because community first responders normally do not step into the response role until the initial crisis has happened and been reported (Gomez & Turoff, 2007; Olaniran & Williams, 2001). Furthermore, a CCC might be able to assess the role of a university medical center during a crisis. For instance, answering the following questions may help such a center implement the AMCM principles in its crisis and disaster planning:

- Does the university hospital have a telemedicine program?
- Can equipment and personnel be deployed using mobile units or devices to help potential patients and victims?
- Is it better to deploy mobile units to assist patients or to have patients transported to the hospital?

These questions can help assist the CCC and rural communities prepare for crises and disasters and, more importantly, address the different needs that may result based on crisis type. After all, central to the AMCM theory and crisis planning is the notion of requisite varieties, which can only be fully engaged or implemented when an organization has fully identified the possible range of crises and responses (Gilpin & Murphy, 2008; Olaniran & Williams, 2001, 2012; Williams & Olaniran, 1998).

Equipment

Having the right equipment for the right job is of essence in crisis communication and management. In evaluation or identification of community resources, making sure the right equipment is readily available in the community is a must. Equipment can include transportation for moving or evacuating individuals from the crisis site to safety. However, for the purpose of this chapter, the category of equipment will primarily mean tools like construction equipment and earth-movers to reach members of the public in case of natural disasters, hurricanes, tornadoes, and mine accidents. How many times have we witnessed the demise of people on our TV screens in mine accidents, and watched the challenges of figuring out how to reach the victims in time to prevent a catastrophe? The case of the Chilean mine resonates in people's memories, as the world watched for days before the right people with the right equipment were able to bring victims to safety. Fortunately, the Chilean mine was a success story, but the West Virginia coal miners' plight was not, and it, in essence, underscored the need to have the right equipment in times of crises. Not only is it important to have the right equipment, but one must also have competent and skilled people to operate such equipment. Otherwise, having the right equipment without the appropriate personnel to operate it is as bad as not having the proper equipment to start with.

For instance, the BP Gulf disaster reminds us of the negative consequences of lack of equipment, and Hurricane Sandy presented the need for and deployment of heavy equipment for rescue and recovery. For instance, military transport planes with more than 600 tons of power-restoration equipment, vehicles, and a crew from Southern California were being sent to the Northeast, along with marines and sailors from Camp Lejeune in North Carolina (DeWind, 2012). At times it is necessary for the nation's military and defense departments to be involved, especially when resources at the local or state levels are inadequate to address a particular crisis (see Figures 3.1 and 3.2). In particular, the Pentagon, in providing help with Hurricane Sandy, airlifted power-restoration experts and trucks across the country, from California to New York, to bolster efforts to assist the millions of people still living in darkness days after the hurricane hit the northeastern United States. The Pentagon brings its own unique skills and equipment to the nation's natural disaster responses, and acting in a timely manner in the wake of calamities (forecast and unexpected) is what the U.S. military specializes in (DeWind, 2012). In addition to having the right equipment with Hurricane Sandy, 12,000 National Guard troops were mobilized to active-duty status in more than a dozen states to provide rescue and relief effort (see Figure 3.2). In essence, it is not just about having the right equipment—the ability to mobilize or to deploy them along with personnel who know how to operate such equipment is imperative. Other equipment may need to be deployed ahead of a crisis (e.g., a bulldozer for sand bags in preparation for a hurricane as indicated in figure 3.3). More important, however, is the community's responsibility to put in place an action plan that involves instructions to access these resources. Access to equipment may call for an advance application or approval process, something that the CCC should spearhead by having a procedure in place for securing access (Williams & Olaniran, 1994). A case in point was the Exxon *Valdez* crisis in Alaska, where Exxon blamed the government of Alaska for not being able to use the chemical dispersant to deal with the oil spill. However, Olaniran and Williams contend that Exxon should have sought preapproval for the use of the dispersant, even before the report of the oil spill, to be able to move in a timely manner to address the crisis.

In this context, industries, organizations, communities, and state and federal agencies must maintain communication that results in positive outcomes or facilitates effective crisis and disaster plans, preparation, and, consequently, crisis management (Olaniran & Williams, 2001, 2012; Seeger, Sellnow, & Ulmer, 2010; Williams & Olaniran, 1998).

Figure 3.1: Air Force crews offload Southern California Edison power repair equipment from a C-5 Galaxy at Stewart Air National Guard Base in Newburgh, NY, November 1, 2012. The Department of Defense initiated the airlift operation to aid recovery efforts in Hurricane Sandy's aftermath. (Photo: USAF; source: Department of Defense).

Figure 3.2: Soldiers assist residents displaced by Hurricane Sandy in Hoboken, NJ, October 31, 2012. The soldiers were assigned to the New Jersey National Guard. (Photo DoD; source: The U.S. Coast Guard).

Figure 3.3: Heavy equipment is used to pile sand into berms before the arrival of Hurricane Sandy.

Communication and Crisis Response Team Interoperability

In order for crisis teams to function effectively, communication coordination and interoperability among people and communication media are critical. It is argued that most crisis incidences require more than one organizational response team (i.e., firefighters or police along with the EMS), but the nature of the crisis may require one agency to play a stronger role in the aid-rendering response effort (Gomez & Turoff, 2007). For example, in case of a fire, the firefighters would bear the brunt of the work. Yet it is equally important for the EMS to be present to lend a helping hand to injured individuals, and police presence may be needed to reroute travelers to alternate directions or to provide measures to control the scene. Most response teams communicate with their own communication devices as approved by their respective agencies. Therefore, the differences in selected communication devices may hinder communication across organizational boundaries and in essence not be interoperational. Gomez and Turoff (2007) alluded to the DHS representative David Boyd's assertion that interoperability is a significant challenge in crisis situations (see also Wood, 2006). Usually, interoperability brings into the forefront the issue of radio or two-way communication, such as by walkie-talkie or other mobile telephony. Within such communication encounters, the need and ability to interact or to communicate effectively are important. This factor is important for both one-way and two-way communication, and hence the ability to collaborate and to coordinate response team activities using or deploying ICTs. Perhaps one of the biggest challenges to effective communication over ICTs is the agency-specific lingo used. Boyd points to exacerbating factors in interoperability, such as the language and culture of the communicator (Turoff & Gomez, 2007). Thus, in an attempt to overcome this challenge, it is important that crisis response teams be prepared for communication difficulty and be proactive about addressing the issue. One way of addressing the problem is by making sure that when individuals are communicating in a collaborative manner with other agencies in the crisis response team, they communicate in plain language.

Times of crises require that one does not leave anything to chance—rescue operations and lives may very well depend on the success or failure to communicate clearly and effectively. Others have suggested fostering ICT e-readiness (i.e., the ability to use ICTs with the ease of use of the communication system) as a way to increase the effectiveness of the technology and to overcome other communication issues associated with any crisis response effort (Gomez & Turoff, 2007; Olaniran, 2011; Olaniran & Williams, 2012; Turoff, Chumer, Van De Walle, & Yao, 2004). Reducing communication challenges and coordination problems will allow crisis response teams to focus on their roles in a given crisis and help circumvent uncertainties emanating from a crisis. Furthermore, it is the responsibility of the CCC to help different institutions, agencies, and people understand the critical nature of their role in crisis preparedness and the compounding consequences of the failure to perform those roles. Communication interaction through ICTs goes beyond the ability to exchange information; it may include other ICT devices, such as operating cameras (photos and videos) and GPS. Consequently, proper training to achieve a certain degree of competence and comfort level with a given ICT tool is called for.

Aligning Emergency Response Events

From a crisis preparedness standpoint, it is important that crisis response teams are able to put in place a crisis plan that allows communities and their residents to receive and understand risk communication messages. Risk communication consists of messages designed with the intent to inform the public about an impending danger or threat and to persuade the target audience to engage in action that will help reduce the risk (Seeger & Reynolds, 2007). Risk communication would allow assembly of emergency kits and plans to limit risks upon crisis and disaster occurrences. Hence, the CCC can help identify, develop, and foster risk communication, and, more importantly, help to frame messages and present them in a manner whereby the targeted audience would be motivated to comply (Gilpin & Murphy, 2008).

Media

For a significant period of time, media reporting on crises has been the subject of debate and attention in many societies. Crisis communication scholars have acknowledged the interdependent and mutual relations between crises and the media (Williams & Olaniran, 1998). The interdependency between the media and crises has also been presented in other publications (e.g., Ma, 2005; Singer & Endreny, 1993; Tumber & Palmer, 2004; Ungar, 1998; Weerakoon, 1993).

Media outlets (i.e., mass and new media) can play, and do play, a key role in crisis prevention and management. Furthermore, it has been established that media reporting is essential to the nature of perception, management, and the creation of crises (Ma, 2005). Therefore, a key component of CCCs is media and media relations, because the media are instrumental in helping the CCC to communicate its messages to various publics and community constituents. Perhaps of concern to CCCs is the trustworthiness of the media in terms of their ability to communicate information in a neutral and nonsensationalizing way. Regardless of perception, trust takes time to develop, and it is not a responsibility of one party. For instance, the CCC must develop relations with all members of the mass media public (e.g., radio, TV, and print) without appearing to be playing favoritism. Of course, in a large community or media market, this may be daunting, but the problem can easily be overcome by targeting major network mass media. At the same time, the media outlets must be able to communicate the message to their audiences in such a way that media consumers and publics can interpret the message to make meaningful or life-saving decisions in a timely manner.

Mass media ought to be an extension of people's *ears* and *eyes* and ought to be able to provide important information for the public in a way that helps them deal with a crisis situation (Ma, 2005). In times of crises, people look to the media outlet to keep them informed and updated so that they can make decisions regarding whether to evacuate or to seek shelter, stop consuming certain products, and so forth. Public and private sectors are also believed to rely on media coverage for crisis management to engage in corrective actions and in decision making (Lee, 2000).

Media outlets and news reporting have been linked to agenda-setting (Singer & Endremy, 1993). In other words, the manner in which a crisis is developed or escalated can be related to the agenda-setting role when media pick and choose the issue to report and how to report it (i.e., framing); this is why some crisis scholars advocate the need for crisis management practitioners to be in control of messages to the public through media outlets as much as possible. One way of doing this is by building relationships with media members prior to the time of the crisis to allow for trust between parties to develop (Leighton, Pellegrino, & Shelton, 2008; Lukaszewski, 2013; Olaniran & Williams, 2001).

Perhaps one of the ways to accomplish this goal is for the CCC to develop specific message content that can be distributed to members of the media. Literature has indicated that the need to design and to customize messages depends on the stages of the crisis. Three types of messages were identified by Sturges (1994):

- Instructing information—which focuses on telling the affected public how they should react in a crisis situation (Barton 1993; Quarantelli, 1988)
- Adjusting information—which focuses on helping members of the public to mentally or psychologically cope with a crisis (Kuklan, 1986; Shrivastava & Mitroff, 1987)
- Internalizing information—which focuses on helping people formulate opinions about the organization (Newsom & Carrell, 1986)

All three types of information are essential and important, but the emphasis from CCCs should be about the first two types of messages. For instance, when a crisis is seen as imminent, the communication to the public should focus on preparing them for a crisis through specific actions to embark on. As a crisis breaks out, the focus of the message needs to shift to messages that that will keep the public out of harm's way or avoid further damages (Lukaszewski, 2013; Sturges, 1994). For instance, when a tornado hit Shawnee, Oklahoma, in 2013, people from neighboring communities wanted to do something; they wanted to go and help restore power and remove debris, but the information released from first responders was to stay away to avoid complicating an already complex situation. Consequently, people are encouraged to donate cash to responding organizations instead of going there.

Social media, which include varieties of technologies such as the Internet, blogs, chat rooms, Twitter, and Facebook, have a role to play in a crisis because they are quick and easy to deploy. One of the essential aspects of crisis management is the need for quick, accurate, and consistent responses (Coombs, 2007; Palenchar & Freberg, 2012). Thus, what social media allow people to do is to have a sense of empowerment and control over how they receive and consume information. Thus, the CCC needs to figure out a way to incorporate social media into their message delivery. This approach will allow the centers to reach new generations of consumers who have their own preferences on how they consume messages. It may also help the center to reach vulnerable populations, including the home-bound and the disabled (Veil, Mitcham, & Sellow, 2012).

Legal aid and counseling

A crisis brings with it a lot of stress and trauma. Therefore, it is essential that individuals be offered a way to cope with the aftermath of a crisis. While the CCC is not targeted to dealing with individual needs on a one-on-one basis, it can play a significant role in helping coordinate where individuals are able to get help that may be tailored to their needs. Thus, having a directory of centers that offer both legal and counseling information within the community is helpful. For instance, in the aftermath of the school shooting at Newtown, Connecticut, students and parents were offered counseling services, some of which were organized by the branch of the American Red Cross and others through volunteer psychiatrists from within that and neighboring communities (Levin, 2013a).

The list or directory can include varieties of professionals and others, such as psychiatrists, psychologists, social workers, and clergy or spiritual counselors, and their job may extend beyond direct counseling. They can offer guidance to parents on how to talk to their children and allow people to express their emotions. Levin (2013b) acknowledged that some of the counselors after the Newtown shooting wanted to create an environment where children could attempt to get back to their normal routines. The quicker the victims are able to address their physical and emotional needs, the quicker they are able to adjust and to move beyond the crisis. Anthony Ng, an emergency psychiatrist who worked with the Newtown survivors, recognized the important role of psychiatry in crisis and disaster aftermath. He suggests that "psychiatric leadership be aware and even be involved in the post disaster financial response, too" (as cited in Levin, 2013b). It may even be necessary to have counseling opportunities for those treating or working with victims of disasters (i.e., post-Aurora experience was helpful for the Boston Marathon ER workers).

Opinion leaders

Although not often discussed in the crisis literature, opinion leaders can play an important role in crisis readiness and management. Therefore, we encourage their use in the CCC and in other crisis situations. Opinion leaders consist of individuals who may not necessarily hold an elected position in a community but have constituents who have come to see them as credible sources of information and perhaps even regard them as culturally similar and having their best interests at heart. These groups of individuals are easily identifiable in communities because they tend to speak for ethnic minorities and advocate for special causes and interests within their communities. For example, in Lubbock, Texas, two individuals easily fulfilled this role: a former councilman, T. J. Patterson, who is also a publisher of a newspaper, the *Southwest Digest*, which caters to the needs of the Lubbock black community; and Christina Martinez-Garcia, a local publisher of a magazine, *Latino Lubbock*, which caters to the Hispanic community.

In the foregoing discussion, we talk about how CCCs need to get their risk communication and messages across to members of the community. Opinion leaders know how best to get that information across and have been shown to be valuable in interpreting the message in a manner that best gains compliance (Olaniran, 1993; Rogers, 2010), which is critical when centers such as CCCs issue warnings and evacuation orders during crises. They know their community well, and they know the best way to reach traditionally hard-to-reach populations (i.e., vulnerable populations) within their respective communities. Just as there is evidence of good coordination and collaboration in mobilizing community resources, there are often more cases of poor coordination and mobilization. Therefore, a CCC needs to identify and to train the identified opinion leaders in the nature of the crisis or disaster and help them develop key information or risk communication messages (see also

Leighton et al., 2008; Lukaszewski, 2013). The cases presented at the end of this chapter illustrate coordination that is not stellar, and thus the ensuing consequences are less desirable.

Conclusion

The CCCs' role is to figure the best way to coordinate individuals, crisis responses, and response teams in a way that prevents crises, prepares communities for crises, and helps deal with the aftereffects of crises. In an attempt to do this, the CCC needs to assess the available resources and people and decide how best to use them. It is also the responsibility of the CCC to evaluate material and human resources that are not available and to figure out a way to get them prior to an impending crisis. A list of government resources available elsewhere, and how they can be readily deployed when needed, will go a long way in this process, given that certain equipment or technical skills to operate them may be too cost-prohibitive for every community to own or even the state to own, but they are nonetheless essential to effective crisis management. Perhaps the CCCs' best asset is the ability to serve the community as an organized source of information and to coordinate different crisis response teams to allow the community to participate in, and to be protected from, a crisis.

For a case study, see this link: http://www.thedenverchannel.com/news/movie-theater-shooting/aurora-response-is-serving-as-textbook-for-emergency-response

References

Armstrong, M., & Taylor, S. (2014). *Armstrong's handbook of human resource management practice*. Philadelphia: Kogan Page.

Barney, J. B. (2012). Purchasing, supply chain management and sustained competitive advantage: The relevance of resource-based theory. *Journal of Supply Chain Management*, 48(2), 3–6.

Barton, L. (1993). *Crisis in organizations: Managing and communicating in the heat of chaos*. Cincinnati: South-Western.

Basso, K. (1996). *Wisdom sits in places*. Albuquerque: University of New Mexico Press.

Bertot, J. C., McDermott, A., Lincoln, R., Real, B., & Peterson. K. (2012). *2011–2012 public library funding and technology access survey: Survey findings and report*. College Park: Information Policy and Access Center, University of Maryland.

Bishop, B. W., & Veil, S. R. (2013). Public libraries as post-crisis information hubs. *Public Library Quarterly*, *32*(1), 33–45.

Coombs, W. T. (2007). Protecting organization reputations during a crisis: The development and application of situational crisis communication theory. *Corporate Reputation Review*, *10*(3), 163–176.

DeWind, D. (2012, November 1). Hurricane Sandy: How our military are responding (Updates). *The moderate voice*. Retrieved November 2, 2012, from http://themoderatevoice.com/166208/hurricane-sandy-how-our-military-are-responding/.

Drabek, T. E., Key, W. H., Erickson, P. E., & Crowe, J. L. (1975). The impact of disaster on kin relationships. *Journal of Marriage and the Family*, *37*(3), 481–94.

Federal Emergency Management Agency (2001, January 31). Can you go it alone for 3 days? http://www.fema.gov/news/newsrelease.fema?id=7591.

Gilpin, D. R., & Murphy, P. J. (2008). *Crisis management in a complex world*. New York: Oxford University Press.

Gomez, E. A., & Turoff, M. (2007). Community crisis response teams: Leveraging local resources through ICT e-readiness. *Proceedings of the 40th Hawaii International Conference on System Sciences*.

Green, S. L., & T. H. Teper. (2006). The importance of disaster planning for the small public library. *Public Library Quarterly*, *25*(3–4), 47–59.

Hearit, K. M. (2006). *Crisis management by apology: Corporate response to allegations of wrongdoing*. New York: Routledge.

Jaeger, P. T., Langa, L. A., McClure, C., & Bertot. J. C. (2006). The 2004 and 2005 Gulf Coast hurricanes: Evolving roles and lessons learned for public libraries in disaster preparedness and community services. *Public Library Quarterly*, *25* (3–4), 199–214.

Jaques, T. (2007). Issue management and crisis management: An integrated, non-linear, relational construct. *Public Relations Review*, *33*(2), 147–57.

Kelley, M. (2011). ALA Midwinter 2011: FEMA recognizes libraries as essential community organizations. Library Journal, January 11. http://www.libraryjournal. com/lj/home/888739-264/ala_midwinter_2011_fema_recognizes.html.csp.

Kuklan, H. (1986). Managing crises: Challenges and complexities. *SAM Advanced Management Journal*, *51*(4), 39–44.

Lee, B. K. (2000). Audience evaluation process in organizational crisis: A study of consumers in Hong Kong. Unpublished doctoral dissertation, Hong Kong Baptist University.

Leighton, N., Pellegrino, S., & Shelton, T. (2008). There is no substitute for media training. In P. Anthonissen (Ed.). *Crisis communication: Practical PR strategies for reputation management & company survival* (pp. 132–141). London: Kogan Page.

Levin, A. (2013a). Volunteer psychiatrists help Newtown cope with tragedy. *Psychiatric News*, 48(2), 1 12. doi:10.1176/appi.pn.2013.1b15.

Levin, A. (2013b). Disaster psychiatrist fills niche in school-shooting aftermath. *Psychiatric News*, 48(3), 99, doi:10.1176/appi.pn.2013.2a7.

Lindell, M. K., & Perry, R. W. (2004). *Communicating environmental risk in multiethnic communities*. Thousand Oaks, CA: Sage.

Lukaszewski, J. E. (2013). *On crisis communication: What your CEO needs to know about reputation, risk and crisis management*. Brookfield, CT: Rothstein.

Ma, R. (2005). Media, crisis, and SARS: An introduction. *Asian Journal of communication, 15*(3), 241–46.

Mandel, L. H., Bishop, B. W., McClure, C. R., Bertot, J. C., & Jaeger, P. T. (2010. Broadband for public libraries: Importance, issues, and research needs. *Government Information Quarterly, 27* (3), 280–91.

Mileti, D., Sorensen, J., Sorensen, B., & Sutton, J. (2002). *Warning America*. Boulder, CO: Natural Hazards Center.

Newsom, D., & Carrell, B. (1986). *Public relations writing: Form & style* (2nd ed.). Belmont, CA: Wadsworth.

Olaniran, B. A. (1993). An integrative approach for managing successful computer-mediated communication technological innovation. *Ohio Speech Journal, 31*, 37–52.

Olaniran, B. A. (2011). Designing functional virtual learning communities using the Bola Ola Method. *Knowledge Management & E-Learning: An International Journal, 3*(4), 697–710.

Olaniran, B. A., & Williams, D. E. (2001). Anticipatory model of crisis management: A vigilant response to technological crises. In R. L. Heath & G. Vasquez (Eds.), *Handbook of public relations* (pp. 487–500). Thousand Oaks, CA: Sage.

Olaniran, B. A., & Williams, D. E. (2004). Burkian counternature and the vigilant response: An anticipatory model of crisis management and technology. In D. Millar & R. Heath (Eds.), *Responding to crisis: A rhetorical approach to crisis communication* (pp. 75–94) Mahwah, NJ: Lawrence Erlbaum.

Olaniran, B. A., & Williams, D. E. (2012). A multicultural perspective on crisis management: Shell & the Ogoni people of Nigeria. In B. Olaniran, D. Williams, & T. Coombs (Eds.), *Pre-crisis planning, communication, and management: Preparing for the inevitable* (pp. 227–48). New York: Peter Lang.

Palen, L., & Liu, S. B. (2007, April). Citizen communications in crisis: Anticipating a future of ICT-supported public participation. In *Proceedings of the SIGCHI Conference on Human Factors in Computing Systems* (pp. 727–36). New York: ACM.

Palenchar, M., & Freberg, K. (2012). Emergency management planning: Risk, crises, issues and social media. In, B. Olaniran, D. Williams, & T. Coombs (Eds.), *Pre-crisis planning, communication, and management: Preparing for the inevitable* (pp. 147–69). New York: Peter Lang.

Quarantelli, E. L. (1988). Disaster crisis management: A summary of research findings. *Journal of Management Studies, 25*(4), 373–85.

Rogers, E. M. (2010). *Diffusion of innovations*. New York: Simon and Schuster.

Scholl, J. C., Williams, D. E., & Olaniran, B. A. (2005). Preparing for terrorism: A rationale for the crisis communication center. In D. O'Hair, R. Heath, & G. Ledlow (Eds.), *Community preparedness and response to terrorism: Communication and the media* (pp. 243–68). Westport, CT: Praeger.

Seeger, M. W., & Reynolds, B. (2007). Crisis communication and the public health: Integrated approaches and new imperatives. In M. W. Seeger, T. L. Sellnow, & R. R. Ulmer (Eds.), *Crisis communication and the public health* (pp. 3–20). Cresskill, NJ: Hampton.

Seeger, M. W., Sellnow, T. L., & Ulmer, R. R. (2010). Expanding the parameters of crisis communication: From chaos to renewal. In R. Heath (Ed.), *Handbook of public relations* (pp. 489–99). Thousand Oaks, CA: Sage.

Seeger, M. W., Sellnow, T. L., & Ulmer, R. R. (2003). *Communication and organizational crisis*. Greenwood Publishing Group.

Shrivastava, P., & Mitroff I. I. (1987). Strategic management of corporate crisis. *Columbia Journal of World Business*, *22*, 5–11.

Singer, E., & Endreny, P. M. (1993). *Reporting on risk: How the mass media portray accidents, diseases, and other hazards*. New York: Russell Sage Foundation.

Stewart, M. (2013). Aurora theater shooting response serves as textbook for E.R. care, aids Boston hospitals in bombings. http://www.thedenverchannel.com/news/movie-theater-shooting/aurora-response-is-serving-as-textbook-for-emergency-response.

Sturges, D. L. (1994). Communicating through crisis: A strategy for organizational survival. *Management Communication Quarterly*, *7*(3), 297–316.

Tumber, H., & Palmer, J. (2004). *Media at war: The Iraq crisis*. London: Sage.

Turoff, M., Chumer, M., Van De Walle, B., & Yao, X. (2004). The design of a Dynamic Emergency Response Management Information System (DERMIS), *Journal of Information Technology Theory and Application*, 5(4), 1–35.

Ungar, S. (1998). Hot crisis and media reassurance: A comparison of emerging diseases and Ebola in Zaire. *British Journal of Sociology*, 49(1), 36–56.

Veil, S. R., & Bishop, B. W. (2014). Opportunities and challenges for public libraries to enhance community resilience. *Risk analysis*, *34*(4), 721–734.

Veil, S. R., Mitcham, C., & Sellnow, T. (2012): Lost in translation: how emergency managers communicate preparedness with vulnerable populations. In B. Olaniran, D. Williams, & T. Coombs (Eds.), *Pre-crisis planning, communication, and management: Preparing for the inevitable* (pp. 201–225). New York: Peter Lang.

Wachtendorf, T. (2004). *Improvising 9/11: Organizational improvisation following the World Trade Center disaster*. Dissertation, University of Delaware Disaster Research Center.

Weerakoon, B. (1993). Introduction. In *Role of media in a national crisis* (pp. xi–xiv). Singapore: Asian Mass Communication Research and Information Centre.

Weick, K. E. (1979). *The social psychology of organizing* (2nd ed.). New York: Random House.

Weick, K. E. (1996). *Sensemaking in organizations*. Thousand Oaks, CA: Sage.

Will, B. H. (2001). The public library as community crisis center. *Library Journal*, 126(20), 75–77.

Williams, D. E., and Olaniran, B. A. (1994). Exxon's decision-making flaws: The hypervigilant response to the Valdez grounding. *Public Relations Review*, *20*, 5–18.

Williams, D. E., & Olaniran, B. A. (1998). Expanding the crisis planning function: Introducing elements of risk communication to crisis communication practice. *Public Relations Review*, *24*, 387–400.

Wood, D. (2006). Opinion: Interoperability issue still bedevils emergency response. *Newhouse News Services*. http://www.jems.com/products/interoperability/articles/17339/, 2006.

Zaremba, A. J. (2010). *Crisis communication: theory and practice*. New York: Routledge.

Snow in Atlanta: A Case Study

- http://www.usatoday.com/story/news/nation/2014/01/30/atlanta-storm-blame/5050875/
- http://www.bbc.co.uk/news/blogs-echochambers-25968133

West Virginia Chemical Spill

- http://www.fema.gov/ar/disaster/3366
- http://www.csmonitor.com/USA/USA-Update/2014/0112/West-Virginia-chemical-spill-More-days-to-come-without-water

· 4 ·

NEW MEDIA TECHNOLOGIES' ROLE IN THE CCC'S CRISIS MANAGEMENT

Chapter Highlights

- Crisis Communication
- Crisis Communication and Communication technologies (e.g., social media): The Push vs. Pull Media
- Environmental Scanning and Social Media

Overview

This chapter reviews the research and theory of mass media, new and social media, and other information communication technologies (ICTs). It also provides guidelines for how CCCs can utilize local media outlets and take advantage of communication technologies to stay connected with the local public.

Introduction

The fact that an organization must respond to a crisis is one of the basic fundamentals in crisis management. However, how organizations respond or communicate with the public or with the stakeholders is what determines whether the crisis management will be effective or not. In consonance with the principles of the AMCM for the CCC's crisis and disaster management, this chapter contends that planning communication messages and strategies is a critical role that any organization, and the CCC in particular, must undertake. According to Lukasewski (2013), an organization needs to respond to crises and communicate energetically, because their reputation is always on the line. Unfortunately, some organizations and their crisis management staff often fail in this important mission, either because they don't care or because they intentionally ignore their turbulent environments and thereby fail to account for crisis-inducing events (Lerbinger, 2012; Lukasewski, 2013). One of the most important crisis-inducing factors concerns technologies and human factors, specifically the role that new forms of communication technologies play in crisis communication processes. At the same time, most organizations do not understand how to use or to take advantage of these new media platforms in crisis and disaster management (Gruber Smerek, Thomas-Hunt, & James, 2015; Ki & Nekmat, 2014).

In essence, if a crisis creates victims, as Lukasewski (2013) contends, then organizations such as the CCC must figure out ways to accept the responsibility of caring for the victims and for helping them minimize the impacts of crises and disasters. Therefore, it is important that even prior to the actual occurrence of crises and disasters, the CCC must anticipate the needs of its publics and develop a way to communicate with community members in a manner that helps address those needs and concerns. This is important because of the need for appropriate, coherent messages and the speed with which messages must be delivered (Jin, Liu, & Austin, 2014). However, one cannot meet the speed criteria of communication that a crisis demands if one has not given any thought to or developed a plan for public and stakeholder information needs. Lukasewski (2013) seems to echo this idea when he argues that hesitant or confused communication sets crisis management up for failure; such ill-advised communication does not meet the crisis speed requirement and, consequently, creates a foundation for a negative perception of the crisis response. Therefore, the goal in this chapter is to examine the research and theory of mass media, new and social media, and other information communication technologies. The chapter also provides guidelines for how CCCs can utilize local media outlets and take advantage of these technologies to stay connected with the local public during crisis and disaster management.

New Media and Their Roles in the CCC for a Changing Landscape of Crisis and Disaster Communication

The dawn of the digital age and associated communication technologies, specifically, computer-mediated communication (CMC) technologies, have drawn significant attention to the role of ICT in crisis communication and disaster management, either in terms of risk communication (to warn) or in crisis and post-crisis efforts. Consequently, the options available for communicating about a crisis are no longer limited to the linear mass media (Bridgeman, 2008; Gruber, Smerek, Thomas-Hunt, & James, 2015, Jost, Banaji, & Nosek, 2004; Stephens & Malone, 2009), where a public relations effort is considerably controlled by organizations and messages are pushed out to the masses without regard to the communication preferences of the audience. However, stakeholders and other members of the public use or are using peer-generated information to compare or to make sense of official and formal information they receive about a crisis. Furthermore, new media users are engaging these technologies to force crisis managers to be transparent and, at times, engaging in some form of new media activism to secure a desired change. While technology is allowing people to connect with others at a faster rate, the idea of people or stakeholders connecting with others to make sense of information and, in particular, crisis messages, predates some of the newer communication technologies (Palen, Vieweg, Liu, and Hughes, 2014). Stakeholders and members of the public are known to use several sources of information, including families, friends, and others, in their decision making about crisis and disaster management (Sorensen, 2000; Sorenson & Sorenson, 2006).

At the same time, ICT has enhanced the public's *information-seeking* and *self-organizing behaviors* to a great extent, especially in allowing two-way informational flow and in increasing social interactions (Ding & Zhang, 2010; Gruber, Smerek, Thomas-Hunt, & James, 2015; Lukasewski, 2013; Palen, Hiltz ,& Liu, 2007; Palenchar & Freberg, 2012). Specifically, Palen, Hiltz, and Liu (2007) suggest that the increased level of social interaction facilitated by these technologies "moved patterns of communication from the ephemeral to the traceable" (p. 468). For specific types of new media see Appendix 1.

From Centralization to Decentralization

In traditional crisis management, mass media are at the center. However, crisis management and crisis communication have moved beyond traditional crisis orientation into a more pragmatic orientation demanded by increased complexities of the modern societies. While crisis management using mass media focuses on controlling information and messages, the changing landscape of the contemporary society has made such control more complex—or, frankly, an impossible task. Societies have become more fluid and demand an increased level of participation in crisis events and in communication.

From a communication perspective, there is a sudden shift in crisis management from what is characterized as a solid society dominated by mass media centralization to that of a liquid society based on increased demands for participation by different groups (i.e., *participatory democracy*), especially in an increasingly networked environment (Falkheimer & Heide, 2012). In essence, traditional crisis communication is considered a linear transmission process in which mass media is used to push out information without distinction in how different groups and/or stakeholders receive, process, and interpret messages. The participant-oriented approach, on the other hand, recognizes distinctions among different groups that are unique, fragmented, and culturally distinctive, and it also requires an organization to develop a more group-appropriate strategy in the crisis management process and communication (Falkheimer & Heide, 2012).

Bauman (2000) extends the thought on the shift from a centralization to a decentralization perspective by distinguishing between what he calls *solid* and *liquid* societies. A solid society embodies the notion of centralization, that is, having a clear role for different actors. As such, mass communication represents the major mode of communication. Liquid society, on the other hand, is characterized by increasing individual choices where people experience increased uncertainties and continually reexamine their relationships and how they interpret information. Therefore, the line of demarcation that exists in a solid society is challenged, if not considered obsolete, in a liquid society. Lull (2007) extends this idea further by examining the notion of two media cultures on a collision course. First, the push media culture, which characterizes the traditional mass media, allows centralized authorities to disseminate information in one or many channels. Within the push media, audiences are target groups separated only through demographic variables; journalists and professional communicators are message senders, target audiences are receivers. More importantly, in times of a crisis, effective communication in this instance focuses on tightly controlled message formulation and information dissemination to the extent that solid society and media pull culture are considered as one (Falkheimer & Heide, 2012). Second, the pull media culture, in contrast to the push culture, focuses on the different needs of niche audiences and considers the old media notion of sender–receiver as inappropriate or obsolete (Falkheimer & Heide, 2012; Lull, 2007). Pull culture, in essence, is personally based, private, and fast-paced and based on the *paradigm of production*, where audiences are fragmented and consumers are actively engaged in production, consumption, and dissemination of communication messages. The pull culture is apparently more popular among the younger generation, which tends to be more technologically conscious and savvy. The reality is that contemporary consumers want to be more involved in the products, services, and organizations they are dealing with by engaging with them through the ability to provide feedback (Mangold & Faulds, 2009). Companies such as Toyota and Ford Motor are increasingly recognizing the importance of such engagement by their customers to the extent that they are using blogs to make such engagement a reality and thus create the opportunity to learn or to know their customers better.

However, rather than take an either/or approach to the push and pull cultures, one should note that there are places for both approaches in crisis management, especially for an effective CCC. Jenkins (2006) believes that we live in a culture of convergence as far as media is concerned, where old and new media have a place and hence can be integrated. Similarly, message content is able to flow across different or multiple media platforms. Thus the participatory nature of pull culture fosters a participant-to-participant rather than an organization-to-participant dynamic. The essence of participant-to-participant communication can be seen in the recognition that everyday citizens are able to capture significant and useful information, and this idea has prompted major traditional media organizations, such as the BBC, to search for social media in disseminating some of their breaking news stories. Similarly, CNN has set up iReport, a division on its website devoted entirely to citizen journalism (Yeomans, 2012).

Information and communication technologies, in general, are changing how organizations respond to disasters. Certain areas of crises that have found significant use with ICTs include the following: emergency medical service (Holzman, 1999); urban search and rescue (Murphy, 2004); emergency evacuation simulation (Johnson, 2005); emergency dispatch work

(Blandford & Wong, 2004; Pettersson, Randall, & Helgeson, 2002); and information systems modeling (Calloway & Keen, 1996; Turoff, Chumer, Van de Walle, & Yao, 2004; Yuan & Deltor, 2005). However, these areas represent formal organizational response to crisis incidence while neglecting the informal aspects. Thus, CCCs need not focus only on the formal application of technologies but can add the informal aspects. The informal aspect primarily describes how members of the public use the same technology in other to get an accurate assessment of the communication process and effectiveness. Specifically, there is the need to focus on how community members generate and interpret information and how this process impacts organizations' actions or responses to crisis situations (Jin, Liu, & Austin, 2011; Palen & Liu, 2007). Again, this dramatic shift puts the onus of effective crisis communication and management squarely on pre-crisis communication, more so than on other crisis stages. Similarly, such an approach shifts the focus in a crisis from an emphasis on the more stable or structured *command and control* response to the more fluid and less stable social dimension of the environment in which a crisis occurs. For instance, the importance of *strategically matching* crisis communication to information forms and sources and to how organizations address crises is considered to be critical to effective crisis management (Jin et al., 2011). Furthermore, the Internet and the proliferation of smart/mobile phones along with social network sites or social media allow people to connect without the restrictions attributed to geographic locations. It is also important to realize that with an increased sharing of messages via social media, messages have the tendency to reach even individuals who may have no interest in a particular crisis. Thus, there is an implication of either (1) creating or expanding new stakeholders or (2) turning issues of minor concern into a full-blown crisis, depending on how recipients of messages respond. In the days leading up to Hurricane Katrina, the Internet allowed some citizens to engage in emergency planning and response activities. After the hurricane hit, an Internet site (www.katrina.com) was found to be useful in locating displaced people, updating information, and providing links to other websites. The website has since evolved to a forum for Hurricane Katrina memorials and general resources for hurricane preparedness (Palen, Hiltz, & Liu, 2007). The 2010 Haitian earthquake saw the use of social media (i.e., wikis) as a platform for sharing and for brokering knowledge (Yates & Pacquette, 2011).

Also in the aftermath of Hurricane Sandy, which wreaked havoc across New York, Long Island, and the New Jersey shore, it was not surprising that people went on Twitter, Facebook, and YouTube to post and to share images and their reflections of the storm's effect and devastation (Yeomans, 2012). Although the traditional media may be a bit complacent about the role social media plays in our connected society, it is hard to deny the fact that the new media platforms can help them to capture information that otherwise beleaguered and understaffed traditional media operations may not be able to capture (Yeomans, 2012). The currency and significance of information that social media allows are only a few of the advantages about how social media can inform and help in crisis situations. For example, Yeomans (2012) offered how crowdsourced crisis mapping sites provided by the likes of Ushahidi and Google Maps, are helping local people and organizations identify communities crying out for relief and how local people made use of a mapping site, Mappler, to provide an up-to-the-minute map of open gas stations during Hurricane Sandy. In other instances, and probably more aligned with the planning and preparation role of CCCs, maps using global satellite images have been used, and are currently being used, to address disease outbreaks such as malaria and Ebola and general public health issues (*PBS Newshour*, 2015).

The participatory culture also embraces the idea of *collective intelligence*—which means that no single person or entity has all of the necessary answers or solutions to a problem; instead, everybody knows something that can be pooled together to arrive at a solution (Jenkins, 2006). The notion of collective intelligence is believed to be best represented in online communities and forums that allow images, videos, texts, and audio data to be hosted. For example, Flickr allows individuals and especially bloggers to connect, and it hosts several images and videos that connect people to different topics (Falkheimer & Heide, 2012). At the same time, participatory culture embraced by online forums also encourages some sense of activism. For instance, through the ensuing collective intelligence fostered in social media and in online forums, one can help jumpstart potential solutions, and it may also force an organization to take certain corrective actions. For example, when the typhoon Haiyan hit the Philippines, more than 10,000 people were feared dead. Through Facebook, individuals (i.e., Filipinos) in the Chicago area were able to stay in touch and inquire about their loved ones back in their native homeland. Facebook pages allow people to see photos of the damage and share them with individuals within the 1,500-member Filipino American Network of Chicago regarding the safety of their loved ones (Thayer, 2013). The images of the devastation caused by the typhoon also spurred members of the network into action regarding the next course of action (e.g., prayers and fundraising) to address the most pressing needs in the affected regions. Although a participant-culture perspective allowed by new media can be tremendously valuable, it needs to be said that the solutions arrived at through the collective intelligence are not always positive. Comments provided via new media and, in particular, via social media, often include praise, criticism, and suggestions. However, the participatory perspective gives participants the sense of inclusion and responsibility and that individuals have been included in the discourse process, or perhaps that they have been heard (i.e., empowerment) (Jin et al., 2014). Weick (1995) addressed this issue in his sensemaking perspective, which he suggested is retrospective. Specifically, he argued that sensemaking is not about finding *the truth*, but instead it is about the process where people attempt to answer the questions regarding *what is going on*. From a related and similar point of view, feedback through new media, when explored as a whole, provides a *sense of community* where honest and open communication is encouraged and individual engagement is fostered (Mangold & Faulds, 2009; see also Gruber, Smerek, Thomas-Hunt, & James, 2015; Jost et al., 2004.

In essence, crisis management entities, and CCCs in particular, can use social media to provide and to collect useful information. For instance, the social media platform can be used to help disseminate information to community members and to affected populations regarding how a crisis effort is being managed. It can help demonstrate to the local people how hard the emergency workers are working and provide other useful information that individuals may seek or want (e.g., available shelters and spaces, closed roads, and detours). Perhaps one could argue that the central or critical role social media aims at is empowering stakeholders and community members at large. As such, the CCC and its social media efforts should be directed at helping local people make sense of the senselessness that accompanies crises and disasters.

A direct implication of the participant culture is one in which organizational centralization of crisis management and information suggests that the provision for clear, concise, and harmonious messages might be difficult to accomplish in participatory environments that increasingly require decentralization of information. Thus, the desire for control of information during crises needs to be tempered with the emerging demands of the new consumer environment and media structure (Falkheimer & Heide, 2012; Jin et al., 2011).

New Media Role and Crisis Communication Centers

Perhaps the most important role that a CCC offers in crisis situations is information—information in the sense of transmission and consumption. After all, crises and disasters are events that disrupt the normal temporal order of activities and events (Palen, Hiltz, & Liu, 2007). However, in a contemporary networked world, the disruption created by crises and disasters is further complicated by ICTs, which produce *new temporality* and social structure (Castells, 2000). According to Palen, Hiltz, and Liu (2007), new media allow individuals to organize themselves "through information tasks—production, sharing, aggregation, and manipulation—people are able to work in advance of conventional forms of news communication" (p. 476). Consequently, the most pressing needs for individuals within a community or societies in general is access to accurate information to help them make sense of a crisis. Individuals, at the same time, however, will also go to tremendous lengths to gather such information and do so before information is provided by traditional news agencies (e.g., press releases). In other words, the speed at which information is released to the public and stakeholders is important. The question, however, concerns the balancing of the speed of information with accuracy, and the ethical appropriateness of such information. In the following section, we discuss the role of new media in CCCs across the crisis communication stages.

ICTs and Pre-crisis Communication

The major function of CCCs is to anticipate potential crises and to safeguard community members against the occurrence of crises or disasters while minimizing the impact of the crises (Olaniran & Williams, 2001; Scholl, Williams, & Olaniran, 2005). Given the focus of pre-crisis on risk communication, then, a CCC's role is to disseminate warnings and communication messages using multiple media (i.e., old and new). Warning messages can then be disseminated via the Internet through the CCC's website or dedicated forums, such as blogs and Facebook pages. After all, individuals are more likely to check these alternate media sources because of the convenience they offer than wait for official press releases. Another mode of communication that we recommend is for CCCs to offer emergency communication alert systems that community members can subscribe to by providing their mobile or phone numbers. Such communication systems are gaining popularity among the school system, especially in the wake of rampant school shootings. The technology allows an organization, with the push of a button, to disseminate necessary information via text or audio to multiple individuals. Because there are those who might be concerned with issues of privacy, the system should be used as a voluntary subscription service, while CCCs must also make efforts to convince the public that their information will be secure and will only be used for the purposes indicated. With social media, information or messages can be shared with large numbers of people with minimal effort.

ICTs may also help CCCs in its crisis preparedness through retrospective sensemaking (Weick, 1995). By retrospective sensemaking, we mean that a CCC can explore how other communities, and its own, have responded in past crises to develop its strategic communication needs. In an earlier chapter, we provided the example of how Massachusetts General Hospital prepared its emergency room (ER) to accommodate the overload created by the Boston Marathon bombing victims, which it did only after learning or looking at what another ER did during the Aurora, Colorado, theater shooting massacre. Other examples include lessons learned from hurricanes, floods, and other crises. In other words, preparation for crises need not only be confined to lessons learned from a particular community. With several online group forums forming after crises or disasters, through data mining an organization charged with crisis management, such as a CCC, can glean valuable information to determine or to assess the most pressing information needs and perhaps how people prefer to get that information. Accordingly, a CCC can then use such information to develop its own appropriate communication messages and transmit them using multiple media platforms. This is in line with the suggestions in the taxonomy of mediated crises, which suggests that organizations should create different web pages and different contents for different stakeholders (Taylor, 2010; Taylor & Kent, 2007).

At the same time, it may not be enough to just inform the public about an impending crisis. It is equally important that a CCC helps in providing how the information or communication messages can be used. That is, it is important to indicate what actions are needed. For instance, if the crisis involves food-product contaminations, certain questions concerning which brand names to avoid or information regarding food preparation would be of value to community members and to the public. In general, planning and preparation for how to communicate with members of the public and the determination of appropriate media to focus on stakeholders' emotions are paramount. Similarly, to address potential negative or *antiorganization information* that might spread through social media, it is also critical in the pre-crisis stage to anticipate such possibilities and to be ahead of the game to avert it (Jin et al., 2011).

ICTs and Communication during Crises

The aim of crisis communication is to help meaningfully explain what has happened and to present what has been done to address the crisis or to help stakeholders move beyond the crisis. In other words, the crisis communication should focus on *instructing information* about *who, where, why, and how* (Coombs, 1999; Taylor, 2010). Therefore, ICTs play a crucial role in providing necessary information and communication to the affected public in the crisis sense making process. As alluded to earlier, from the onset of a crisis, individuals and groups take to the social media to discuss and to share new information about what happened because they want the opportunity to share information with others. At the same time, individuals are able to provide their own take or reaction to messages. The collective intelligence, as well as opinions and information via these new media, often forces organizations to take certain actions, but in general it forces organizations to respond in one way or the other. Furthermore, ICTs allow individuals in the participatory culture to become information co-producers by sharing their views on the crisis (Falkheimer & Heide, 2012; Stephens & Malone, 2009; Taylor, 2010).

From the CCC standpoint, the very essence of the public's demand for information puts emphasis on the speed with which community members must be notified. It is imperative that CCC personnel be diligent in releasing such information and, again, in utilizing the various ICTs that the public are accustomed to. Certain information and actionable activities can be provided through the organization's websites and links to other resources. At the same time, the automated response notification systems might be important in alerting stakeholders about the crisis while providing them with directions to the dedicated website for more information. Finally, the convergence of the old or traditional radio and TV broadcasts with new media is essential to ensure that most, if not all, individuals have the opportunity of being informed. The important thing is to build into the crisis communication process a multiple layer of redundancy using the available technologies at their disposal. Furthermore, this approach is critical for CCCs because, as individuals gathered in their respective forums to share and evaluate a CCC's action, one of the questions the public will want to address is whether CCCs make reasonable efforts to reach and to inform the public.

ICT usage in crisis communication comes with a major challenge, especially as organizations struggle to balance speed with appropriateness. For instance, Palen, Hiltz, and Liu (2007) discuss how certain rules that work with traditional media when it comes to information notification do not work well with new media. For example, they cited the Virginia Tech shooting as a case that illustrates that the rule of law and ethics guide how information about fatalities can be reported. The law, in particular, calls for the releasing of information to the public only after a next of kin has been notified. Palen, Hiltz, and Liu (2007) suggested that such a rule creates conditions of perceived absence of information, because the same rule doesn't apply to the Internet, where such information can be released prior to next-of-kin notification. Thus, it is no surprise that in times of crises some stakeholders monitor Internet forums and social media to determine whether names of their loved ones appear as possible victims rather than waiting for official notification.

CCCs may not be charged with direct notification of individual victims, but they can at least use the Internet to guide the public to such information. The primary responsibility of CCCs is to help mediate the intensity of the crisis during this stage. Thus, efforts regarding emergency response and assistance should be of primary concern. For instance, the website or dedicated Facebook pages should be used to share videos and pictures of emergency workers or responses to the disaster and crisis. Available media and ICTs can also be used to provide other important information, such as where individuals can seek help, road closures, detours, and temporary shelters. Tools such as Google Maps may be of use to the public during crisis disaster management.

Just as it is important to pay attention to online forums and discussion groups during the pre-crisis stage, it is equally important to do the same during a crisis. For instance, the participatory culture allows diverse individuals and groups to create their own forums and opinions (Falkheimer & Heide, 2012; Jin et al., 2011). Given the diversity, it is possible that different cultural groups have different viewpoints about the crisis and how it is being handled. Attention to this information will help the CCC to assess efforts and to create immediate and appropriate responses that address different stakeholders' concerns. This is something that ICTs offer in a participatory culture that is difficult to do in the centralized culture, where an organization has to wait until the crisis is over, and, by then, there is not much it can do. In essence, ICTs, when used appropriately in times of crisis or disaster, can foster simultaneous responses that are customized to the needs of different stakeholders.

Forms or Types of New Media

Certain types of new media are more prevalent among organizations, depending on the industry. For instance, there are social media and social networking sites (SNS) tailored to marketing and promotion (e.g., Mangold & Faulds, 2009), and some cater to travel-related issues (e.g., Xiang & Gretzel, 2010). Here we provide examples and types of social media and SNSs, along with what they can be used for (see Appendix 1). Four types of social media have been identified as commonly used among online entities, including LinkedIn, Facebook, Twitter, and blogs (Gordon, 2009). According to Yang and Kang (2009), blogs have been found to be essential to organization communication strategy in terms of initiating and managing public discourse (see also Mangold & Faulds, 2009; Palenchar & Freberg, 2012). Social media, in general, are helpful in establishing credibility for a brand and organizational reputation among its key publics (Prentice & Huffman, 2008). Social media also provide an avenue for collecting or for accessing *unfiltered* views of consumers and the public in order for an organization to see how those views might impact it and its future (Marken, 2007; Palenchar & Freberg, 2012).

Perhaps it is worth mentioning that social media are often used synonymously with the social networking sites. SNSs, are, however, limited to the idea of web-based connections that allow individuals to be able to connect and to develop relationships. More specifically, SNSs can be distinguished from other forms of technologies in the sense that the platforms are used to connect with people with whom someone already has existing offline relationships instead of trying to build a new relationship with people one does not know (Ellison, Steinfield, & Lampe, 2007).

Technology in Pre-crisis Communication

The development of modern and new technologies is changing the face of crisis management. Consequently, these newer technologies are emphasizing the need for, and the importance of, pre-crisis communication. However, the impact of technologies should be tempered not so much in adapting or adopting newer technologies for the sake of technology, as these technologies should be deployed strategically for the purpose of improving communication effectiveness between crisis organizations and their key publics. In essence, organizations need to include newer technologies (e.g., social media and SNSs) that aim at improving traditional pre-crisis communication strategies, such as press releases and statements that emphasize adequate crisis planning and preparedness with the public.

We stressed earlier that people look for ways to engage in dialogue and to share information with people in their networks. Social media and SNSs offer the potential for accomplishing this goal. Therefore, paying attention to this need can allow organizations to deploy new media in a way that new media helps stakeholders meet their expectations while allowing organizations to adapt their crisis communication messages in the most effective way (Veil, Buehner, & Palenchar, 2011). It has been suggested that crisis professionals need not *reinvent* but rather *adapt* technologies to the likes of social media to convey what they already know using the new media (Semple, 2009). For example, warnings and risk communication messages can be disseminated to publics using social media in an attempt to help key stakeholders take appropriate measures in preventing a crisis or in reducing crisis impacts. Furthermore, social media can be used to foster the personal touch and provide people with actionable steps in order to make crisis communication messages more effective (Dillard, Shen, & Vail, 2007; Heath, Lee, & Ni, 2009; Palenchar & Freberg, 2012).

Palen, Hiltz, and Liu (2007) offered an example of technology use in pre-crisis or crisis preparedness against Avian flu (H5N1) or influenza A (H1N1), the latter a strain of the Spanish flu pandemic of 1918–19. In preparation for another episode of such a pandemic, Palen, Hiltz, and Liu (2007) recalled citizens taking to online forums to create what is called FluWiki (www.fluwicki.com), which was an attempt to circumvent federal and state agency preparation for local preparedness. The FluWiki was used to help people collaborate on the topic and evolution of the influenza strain by discussing national and international plans, personal and family preparedness, along with documentation of rumors, within an online forum. This effort was hailed for people's attempt to take responsibility for their own and familial safety, especially in the aftermath of the government's mishandling of the Hurricane Katrina crisis (Palen, Hiltz, & Liu, 2007). As such, technologies offer individuals participating in the FluWiki the opportunity to empower themselves while adhering to the principle that successful disaster preparedness is based on specific action plans made by individual households. A case in point involved health officials in Houston, Texas, who reported deaths from H1N1 in the area despite the fact that the 2013 flu vaccine included the H1N1 virus strain to prevent people from contracting the virus as part of the lesson learned after the 2009 outbreak (KVUE.com).

Hence, it is the responsibility of organizations such as the CCC to stay on top of potential threats to their communities and then communicate risk and available information so that publics can protect themselves. CCCs can also point individuals to the web link for the Center for Disease Control (CDC), where individuals can do their own due diligence or decide if they want to organize their own local online forum about such threats or to connect with an already existing nationally functioning forum. Technologies, in essence, offer the opportunity to make such plans or action steps a possible reality. Although sentiment measurement—a computer analytical way of gauging emotion (i.e., negative)—focuses on a crisis and on a post-crisis (Lukasewski, 2013), it is believed that sentiment measurement can also be of value in a pre-crisis by using it to learn about stakeholders' reactions to issues that can potentially develop into a full-blown crisis. Thus, CCCs can use such sentiment or information to help develop a crisis and disaster prevention or response plan. In a way, social media provides ways for an organization to stay on top of what people are saying (Lukasewski, 2013; Veil, Buehner, & Palenchar, 2011; Wright & Hinson, 2009)

To this end, the new media provides valuable tools that can assist in identifying potential sources of crises and threats at an early stage of development and to determine where intervention is called for. Thus, an organization may create early warning systems or parameters, such as a Google search or an alert system that provides prompt notification. For example, it has been suggested that a search system can monitor organization environment or target a high-profile individual associated with a corporation in social media contexts, for example, by using Samepoint or Twitter, on a regular basis to determine how what is being discussed could affect an organization or set a course for a potential crisis (Lukasewski, 2013; Palenchar & Freberg, 2012). Furthermore, monitoring how activist groups communicate about an organization within the social media is another area that can sensitize CCCs and serve as a precursor to an early onset of a potential crisis (e.g., Han & Zhang, 2009). In other words, *environmental scanning* is essential to identifying threats to an organization or to a community, and social media tools can help to obtain the necessary information for gathering data about potential crises.

As increased use of technologies becomes the norm, crisis professionals are warned to safeguard against misuse of these technologies during pre-crisis communication. For example, it is suggested that crisis practitioners be aware and safeguard against the potential of giving stakeholders incorrect information or allowing rumors about a crisis event to fester (Palenchar & Freberg, 2012; Vieweg, Palen, Liu, Hughes, & Sutton, 2008).

New Media Use during Crises

The prevalence of mobile telephony has also brought their applications to center stage in crisis communication and has aided social media capabilities. We are starting to see the use of text messages and Twitter via mobile phones as a way to communicate about crises, as in the Arapahoe High School shooting incident. In one occasion, a father and son engaged in texting to give accounts of what happened at the school and informed one another of the steps being taken to safeguard the students even when the school was in lockdown. Complete text messages can be found at the following web link: http://alternatekev.me/2013/12/16/texts-from-my-son-during-the-arapahoe-shooting/.

Later, the father uploaded the transcript of his interactions with his son online to websites that allowed other members of the public to offer comments regarding the incident. In just a few hours, the post generated about 147 comments. In the same crisis, other parents were calling and texting to check on their children after they heard about the shooting (Steffen, 2013).

The high school in question also used social media to disseminate information regarding the school schedule for the remainder of the semester by posting the schedule and notes to the entire community on Scribd (see Appendix 2). Scribd is an online information repository that allows individuals to easily access readable information from any web-enabled mobile device, such as tablets and smartphones. It also allows individuals to upload, download, print, and share content.

The goal of communication during a crisis is to provide support while allowing people the opportunity and the forum to cope or to deal with their concerns. Social media and other technologies facilitate such forums in addition to facilitating coping mechanisms. After all, when a crisis happens, people almost always wonder and ask themselves why. More appropriately, they ask about what they could have done differently, either to prevent the crisis from happening or just to deal with the varieties of emotions they are dealing with. It helps when an organization sends out, via varieties of new media, messages of reassurance that it is doing everything it can to address stakeholders' needs. Perhaps it is important to be reminded that organizations like CCCs are not supposed to have all of the answers or solutions but should be helpful in pointing people to places where they might be able to get some of those answers. For example, providing the public with a place where they can receive good counseling is appropriate. Furthermore, organizations dealing with crises may be able to benefit from the public in terms of accessing key information from the public's use of communication technologies. For instance, society has come to rely on eyewitnesses' accounts and documentaries of events simply because they were in the right place at the right time (Palen, Hiltz, & Liu, 2007). Thus, individuals equipped with cell phones and video cameras can document crises and emergencies and then share the information over the Internet, which was the case during the 9/11 terrorist attacks. Similarly, with nearly everyone equipped with cell phones and GPS-enabled camera phones, it is believed to be easier than ever for bystanders to obtain and to disseminate crisis information in what has been termed *citizens-as-participants* in managing crises (Palen, Hiltz, & Liu, 2007).

Another example is the accruing benefits from multimodal ways of communicating and gathering information during a crisis for crisis management organizations, such as a CCC, as was evident following the 2009 H1N1 flu outbreak. For instance, the CDC posted guidance for schools recommending school closures upon a suspected or actual case of H1N1 flu (CDC, 2012). This guidance was provided to lessen the risk of spreading the virus within communities. However, more information was presented that indicated a lower risk of severe illness and death from H1N1, along with feedback from parents about the undue burden created by the CDC recommendation that parents arrange child care or stay at home with their children. The CDC, in response to the new information, changed its recommendation within six days of the first notice (CDC, 2012).

All in all, the linear depiction of information dissemination about a crisis from crisis managers to traditional news media is considered slow and outdated (Mileti, Sorensen, Sorensen, & Sutton, 2002). But in a networked world with social media and other online forums, people who are geographically removed from crises and disasters are still able to participate and to provide valuable assistance (Hiltz & Turoff, 1978; Palen, Hiltz, & Liu, 2007). At the same time, individuals enabled through online participation are able to learn from personal stories and from the information-sharing process.

With increased use of modern technologies and social media, organizations like the CCC must operate around the clock with a sense of urgency because the public demands immediate access to information. However, when organizations fail to provide the necessary information during a crisis, these new media platforms create a way for people to fabricate their own information or perpetuate rumors. Coombs (2007) summed it by indicating that effective crisis communication must be quick, accurate, and consistent, and he also advises the need for crisis communication to stress public safety issues and concerns. Reynolds and Seeger (2005) also suggest that crisis communication tries to explain the crisis event and provides likely outcomes and *harm-reducing information* to the affected public. At the same time, crisis managers must strive to reduce the risk of information overload during a crisis. Some researchers have argued for the provision of information on a need-to-know basis, while others have suggested taking a more proactive approach—where specific information designed to reduce victims' uncertainties takes precedence over general information and is communicated to the targeted audience directly (Cordner & Scarborough, 2010; Olaniran & Williams, 2001; Olaniran, Scholl, Williams, & Boyer, 2011). From this standpoint, the CDC was able to employ computer modeling to anticipate the potential or grave danger that the Ebola crisis of 2014 posed to the world. From the modeling, the CDC was able to put in a measure, including behavioral changes, such as hand washing and refraining from ritual body washing of the dead prior to burial, to stem Ebola from escalating to a worldwide epidemic (*PBS Newshour*, 2015). Other general information can be provided through an organization's blogs or through dedicated websites.

For crisis management organizations such as CCCs, technologies and, in particular, social media, can be used to set in motion activities from established partnerships put in place prior to the onset of a crisis or disaster. For example, the immediacy offered by communication technologies and deployment of social media is crucial to the coordination of messages among relevant organizations. As a case in point, Avery and Kim (2009) found in their study of press releases about Avian flu that public information officers were proactive in coordinating their efforts with other partnership organizations, such as government agencies, in an attempt to disseminate accurate information. The H1N1 outbreak in Houston, Texas, also depicts the urgency of such coordination among partner organizations. It was reported that local-area hospital officials spent an afternoon on conference calls comparing notes from all of the cases in an attempt to determine symptoms and to evaluate an effective treatment plan (Desel, 2013). The collected information will then be shared publicly with other local health officials and with the Centers for Disease Control and Prevention along with the Texas Department of Health.

At the same time, Palenchar and Freberg (2012) cautioned that the use of social and mobile communication media with such coordination requires strong partnerships. This is in line with the tenets of the anticipatory model of crisis communication, which emphasizes the need to build or to put in place partnerships prior to the onset of a crisis, with the understanding that crisis planning and crisis management go hand in hand with communication during a crisis and consequently impact the degree of effectiveness with a particular crisis (Lukasewski, 2013; Olaniran & Williams, 2012; Scholl, Williams, & Olaniran, 2005).

Similarly, social media can be deployed by CCCs to alert community members to available resources during disasters and crises. Social media tools can also be used to provide up-to-date information regarding road diversion and alternate routes in case of chemical spills and other disasters that may impact the public. For example, the CDC used its Facebook page to provide updates on seasonal flu information (CDC, 2012). The social media tools can also be used to alert hospitals and clinics regarding whether to mobilize their respective emergency and crisis response plans.

It needs to be mentioned that inclusion of public participation in the CCC's organizational management of disasters involves more than handling erroneous information or messages. Quite the contrary, the process requires an acknowledgment and appreciation from the onset of crises and emergencies that the public has a primary role in crisis response (Veil, Buehner, & Palenchar; 2011; Wright & Hinson, 2009). ICT use, however, to a great extent only demonstrates to a wider audience the kind of role the public already plays in their societies during and after crisis events. To this end, Palen, Hiltz, and Liu (2007) warned that existing political trends in the United States that favor *command-and-control-style* crisis management and neglect to appreciate the role the public plays as "first responders" and key players in crisis situations risk the failure to include citizen activity and citizen-generated information in the key warning, response, and relief efforts (see also, Lukasewski, 2013; *PBS Newshour*, 2015; Veil, Buehner, & Palenchar, 2011).

New Media Use in Post-crisis Communication

The use of new media and, in particular, social media, can involve a two-pronged approach in post-crisis communication. First, the media can be used to reinforce risk communication that enables individuals to move beyond a particular crisis or disaster. Second, however, and perhaps the primary emphasis of post-crisis communication, is in evaluating the effectiveness of crisis management and communication. Therefore, it is paramount that organizations such as the CCC evaluate its choice of communication media and messages and how such messages were received or helped in alleviating a particular crisis. For instance, a CCC might want to answer the following questions:

- To what degree did the choice of communication media allow community members to receive both risk and crisis messages in a timely and effective manner?
- To what extent were at-risk or vulnerable populations aided by the new media?
- Were the crisis management or communication strategies well received and helpful?

If the answers to any of these questions receive less than positive affirmation, then it is imperative that the CCC figure out what went wrong and how the problem could have been corrected. This is important as the CCC and members of the community move beyond a crisis or disaster and prepare for similar or other crises in the future. After all, the goal of any crisis management is not only to overcome a particular crisis but also to provide the necessary information that helps stakeholders feel empowered and be crisis-ready in the future. One area to examine in post-crisis communication with the new media is the discrepancies among given community members in their access to new communication technologies, which is complicated with technology literacy, especially among minority groups and other vulnerable populations. Disparity in accessibility to certain communication media can impact the timeliness of messages along with how risk messages were received to take appropriate actions that can help overcome crises or to cope with a given disaster. Other potential areas ripe for post-crisis communication lie in the evaluation of how certain messages were received by different members of the communities. For example:

- Are there certain cultural factors or beliefs that hinder the reception of communication messages?
- How can they be overcome?
- Would certain communication technology or social media be more effective than the chosen one in disseminating crisis communication and messages?
- Were assumptions and choices made with regard to technology and crisis communication considered ethical and appropriate to engage in risk-reducing behavior?
- Answers to each of these questions are germane to proper functioning of a CCC.

Conclusion

New communication technology media serve as valuable tools in crisis and disaster management in any environment. However, communication scholars and other crisis management practitioners need to remember that they are tools and not replacements for good strategic decision making in the course of a crisis or disaster. This idea was emphasized by Kent (2010), who suggested that social media use in reaching the publics need not be mistaken or confused with the need to develop and to maintain viable organization narratives. Kent further suggested that there is no substitute for strengthening organization and public relationships with identification and persuasion, and we should recognize that social media, in particular, helps nurture organizational goals and are viable replacements for more expensive traditional media. Social media and other new communication technology media have a place, an important place in crisis management. However, these media must be used in a manner that helps accomplish organizational goals by recognizing that crisis management professionals have unique skills that they bring into crisis situations and disaster management, which no communication media can replace.

Appendix 1: Types of Social Media and Uses

SNS: Meetup, Myspace, Facebook, and Faceparty
Video sharing: YouTube
Photo sharing: Flickr

Music sharing: Jamendo
Content sharing: Piczo
Blogs: Blogspot, Cnet, and others company sponsored
Collaborative websites: Wikipedia, Second Life, and GoToMeeting
Social bookmarking sites: Digg, del.icio.us, Newsvine, Mixx it, and Reddit
News and Trending Topics: Topix
For a more comprehensive list of SNSs and social media, see also:

Mangold, W. G., & Faulds, D. J. (2009). Social media: The new hybrid element of the promotion mix. *Business Horizons, 52*(4), 357–65.
Xiang, Z., & Gretzel, U. (2010). Role of social media in online travel information search. *Tourism Management, 31*(2), 179–88.

Appendix 2: Arapahoe High School Schedule Posted by Michael Lee Roberts on Scribd

(http://www.scribd.com/doc/191781793/Arapahoe-High-School-Schedule)

Dear Arapahoe Community, December 15, 2013

Thank you for your continued support as we work through the details of the upcoming week. The tentative schedule for the week is outlined below. Some items remain dependent on the availability of our building. Information regarding finals and events following the winter break will be communicated later this week.

Due to logistics such as parking, pickup, and drop-off, we will be staggering the reentry and pickup of belongings for staff and students. This structure will also offer an opportunity for our teachers to reconnect with kids in a smaller environment,

The students' belongings are located in their five-hour MWF classroom. Therefore, we are asking students to report to those classrooms at the start times indicated below. Teachers will be in the room when students arrive. They will be given time to check in, talk, and debrief with their teacher. We have provided a two-hour window, but there is no obligation for students to remain the entire time. We will have mental health staff available in the cafeteria for any students needing to take advantage of these services. We are asking that students arrive at the beginning of their designated time period, and leave AHS (Arapahoe High School) at the end of their scheduled time period so that we can prepare for the arrival of the next class. Siblings in different grade levels will be accommodated to help you avoid multiple trips to AHS.

The Arapahoe faculty and staff, in addition to mental health workers, will be in the building throughout the week. We are encouraging parents to allow students the opportunity to enter AHS independently. We appreciate your trust and cooperation on this. Again, Warriors, please continue to take care of one another!! Stay

WARRIOR STRONG!!!

Natalie Pramenko
Arapahoe High School Principal

TENTATIVE SCHEDULE FOR THE WEEK OF DECEMBER 16 THROUGH DECEMBER 20

Monday—NO ACCESS to AHS
Arapahoe High School remains closed
LPS Counseling Center at Shepherd of the Hills is available to all member of the Arapahoe community as needed.

9:00 a.m.–2:00 p.m.
Please check the LPS Website for updates as to further mental health support throughout the week.

Tuesday—NO ACCESS to AHS
Arapahoe High School remains closed.

Wednesday—NO STUDENT ACCESS to AHS
Faculty return to Arapahoe High School.
Faculty and staff will be able to retrieve all belongings left in the building.
Toys for Tots distribution will continue at Valverde Elementary School for those students signed up to participate.
Please check the LPS Website pickup time and location.

Thursday—UPPER-CLASS STUDENT ACCESS to AHS
Seniors Class of 2014—8:00 a.m.–10:00 a.m.
Juniors Class of 2015—11:00 a.m.–1:00 p.m.

Friday—LOWER-CLASS STUDENT ACCESS to AHS
Sophomores Class of 2016—8:00 a.m.–10:00 a.m.
Freshmen Class of 2017—11:00 a.m.–1:00 p.m.

References

Avery, E. J., & Kim, S. (2009). Anticipating or precipitating crisis? Health agencies may not be heeding best practice advice in avian flu press releases. *Journal of Public Relations Research, 21*(2), 187–197.

Bauman, Z. (2000). *Liquid modernity*. Malden, MA: Polity Press.

Blandford, A., & William Wong, B. L. (2004). Situation awareness in emergency medical dispatch. *International Journal of Human-Computer Studies, 61*(4), 421–52.

Bridgeman, R. (2008). Crisis communication and the net. *Crisis communication: practical PR strategies for reputation management and company survival* (pp. 169–177). Philadelphia: Kogan Page.

Calloway, L. J., & Keen, P. (1996). Organizing for crisis response. *Journal of Information Technology, 11*(1), 13–27.

Castells, M. (2000). *The rise of the network society: The information age: Economy, society and culture*. Vol 1. (2nd ed.). Malden, MA: Blackwell.

CDC (Centers for Disease Control). (2012). *Crisis, emergencies, and risk communication (CERC)* (12th ed.). http://emergency.cdc.gov/cerc/pdf/CERC_2012edition.pdf.

Coombs, W. T. (2007). Protecting organization reputations during a crisis: The development and application of situational crisis communication theory. *Corporate Reputation Review, 10*, 163–76.

Coombs, W. T. (1999). Information and compassion in crisis responses: A test of their effects. *Journal of public relations research, 11*(2), 125–142.

Cordner, G., & Scarborough, K. (2010, January). Information sharing: Exploring the intersection of policing with national and military intelligence. *Homeland Security Affairs*, 6(1), 78–96.

Desel, J. (2013, December 19). H1N1 kills 6 people, leaves 14 critically ill in Greater Houston area. KHOU.com. http://www.khou.com/news/local/BREAKING-Houston-area-health-officials-say-6-deaths-linked-to-H1N1-236620791.html.

Dillard, J. P., Shen, L., & Vail, R. G. (2007). Does perceived message effectiveness cause persuasion or vice versa? 17 consistent answers. *Human Communication Research, 33*, 467–88.

Ding, H., & Zhang, J. (2010). Social media and participatory risk communication during the H1N1 flu epidemic: A comparative study. *China Media Research*, 6, 80–90.

Ellison, N. B., Steinfield, C., & Lampe, C. (2007). The benefits of Facebook "friends": Social capital and college students' use of online social network sites. *Journal of Computer-Mediated Communication, 12*, 1143–68.

Falkheimer, J., & Heide, M. (2012). Participatory pre-crisis and crisis communication: A conceptual approach. In B. Olaniran, D. Williams, & W. T. Coombs (Eds.), *Pre-crisis planning, communication and management: Preparing for the inevitable* (pp. 37–56). New York: Peter Lang.

Gordon, J. (2009). The coming change in social media business applications. White paper for *Social Media Today*. http://www.socialscoop.biz/SMT_ whitepaper_biz.pdf.

Gruber, D. A., Smerek, R. E., Thomas-Hunt, M. C., & James, E. H. (2015). The real-time power of Twitter: Crisis management and leadership in an age of social media. *Business Horizons*, 58(2), 172–82.

Jost, J. T., Banaji, M. R., & Nosek, B. A. (2004). A decade of system justification theory: Accumulated evidence of conscious and unconscious bolstering of the status quo. *Political Psychology*, *25* (6), 881–919.

Han, G. K., & Zhang, A. (2009). Starbucks is forbidden in the Forbidden City: Blog, circuit of culture and informal public relations campaign in China. *Public Relations Review*, *35*(4), 395–401.

Heath, R. L., Lee, J., & Ni, L. (2009). Crisis and risk approaches to emergency management planning and communication: The role of similarity and sensitivity. *Journal of Public Relations Research*, *21*(2), 123–41.

Hiltz, S. R., & Turoff, M. (1978). *The network nation: Human communication via computer*. Reading, MA: Addison-Wesley.

Holzman, T. G. (1999). Computer-human interface solutions for emergency medical care. *Interactions*, 6(3), 13–24.

Jenkins, H. (2006). *Convergence culture. Where old and new media collide*. New York: New York University Press.

Jin, Y., Liu, B. F., & Austin, L. L. (2014). Examining the role of social media in effective crisis management: The effects of crisis origin, information form, and source on publics' crisis responses. *Communication Research*, *41*, 74–94.

Johnson, C. W. (2005). Applying the lessons of the attack on the World Trade Center, 11th September 2001, to the design & use of interactive evacuation simulations. *Proceedings of CHI 2005*. Portland, Oregon, USA, ACM, 651–660.

Jost, J. T., Banaji, M. R., & Nosek, B. A. (2004). A decade of system justification theory: Accumulated evidence of conscious and unconscious bolstering of the status quo. *Political psychology*, *25*(6), 881–919.

Kent, M. L. (2010). Directions in social media for professionals and scholars. In R. Heath (Ed.), *Sage handbook of public relations* (643–656). Thousand Oaks, CA: Sage.

Ki, E. J., & Nekmat, E. (2014). Situational crisis communication and interactivity: Usage and effectiveness of Facebook for crisis management by Fortune 500 companies. *Computers in Human Behavior*, *35*, 140–47.

KVUE.com (2013, December 19). Health officials say 6 deaths linked to H1N. Retrieved from http://www.kvue.com/news/Health-officials-say-6-deaths-linked-to-H1N1-236622341.html.

Lerbinger, O. (2012). *The crisis manager*. New York: Routledge.

Lukasewski, J. E. (2013). *On crisis communication: What your CEO needs to know about reputation, risk and crisis management*. Brookfield, CT: Rothstein.

Lull, J. (2007). The push and pull of culture. *Culture-on-demand: Communication in a crisis world*. Malden, MA: Blackwell.

Mangold, W. G., & Faulds, D. J. (2009). Social media: The new hybrid element of the promotion mix. *Business Horizons*, 52, 357–65.

Marken, G. A. (2007). Social media ... The hunted can become the hunter. *Public Relations Quarterly*, 52(4), 9–12.

Mileti, D., Sorensen, J., Sorensen, B. V., & Sutton, J. (2002). *Warning America*. Natural Hazards Research and Applications Information Center, Boulder, CO.

Murphy, R. (2004). Human-robot interaction in rescue robotics. *IEEE Transactions on Systems, Man, & Cybernetics*, *32*(2), 138–53.

Olaniran, B. A., Scholl, J., Williams, D. E., & Boyer, L. (2011). Johnson and Johnson phantom recall: A fall from grace or a re-visit of the ghost of the past. *Public Relations Review*. doi:10.1016/j.pubrev.2011.08.001.

Olaniran, B. A., & Williams, D. E. (2001). Anticipatory model of crisis management: A vigilant response to technological crises. In R. L. Heath (Ed.), *Handbook of public relations* (pp. 487–500).Thousand Oaks, CA: Sage.

Olaniran, B. A., & Williams, D. E. (2012). The need for an anticipatory perspective in crisis communication. In B. Olaniran, D. Williams, & W. Coombs (Eds.), *Pre-crisis planning, communication and management: Preparing for the inevitable* (pp. 3–16). New York: Peter Lang.

PBS Newshour (2015, April, 24). http://www.pbs.org/newshour/bb/maps-packed-data-help-scientists-fight-malaria/.

PBS Newshour (2015, June 11). Using the power of prediction to halt Ebola in its tracks. *PBS Newshour*. From http://www.pbs.org/newshour/bb/using-power-prediction-halt-ebola-tracks/#transcript.

Palen, L., Hiltz, S. R., & Liu, S. B. (2007, March). Online forums supporting grassroots participation in emergency preparedness and response. *Communications of the ACM*, 50(3), 54–58.

Palen, L., & Liu, S. B. (2007, April). Citizen communications in crisis: Anticipating a future of ICT-supported public participation. In *Proceedings of the SIGCHI Conference on Human Factors in Computing Systems* (pp. 727–36). New York: ACM.

Palen, L., Vieweg, S., Liu, S. B., & Hughes, A. L. (2014). Crisis in a networked world features of computer-mediated communication in the April 16, 2007, Virginia Tech Event. *Social Science Computer Review*, *27*(4), 467–80.

Palenchar, M., & Freberg, K. (2012). Emergency management planning: Risk, crises, issues, and social media. In B. Olaniran, D. Williams, & W. Coombs (Eds.), *Pre-crisis planning, communication and management: Preparing for the inevitable* (pp. 147–70). New York: Peter Lang.

Pettersson, M., Randall, D., & Helgeson, B. (2002). Ambiguities, awareness and economy: A study of emergency service work. In *Proceedings of the 2002 ACM conference on Computer supported cooperative work* (pp. 286–295).

Prentice, S., & Huffman, E. (2008, March). Social media's new role in emergency management. *Idaho National Laboratory*. *Proceedings from Emergency Management and Robotics for Hazardous Environments*. Albuquerque, New Mexico. March 9–12, 2008.

Reynolds, B., & Seeger, M. W. (2005). Crisis and emergency risk communication as an integrative model. *Journal of Health Communication*, *10*, 43–55.

Scholl, J., Williams, D., & Olaniran, B (2005). Preparing for terrorism: A rationale for crisis communication center. In H. O'Hair, R. Heath., & G. Ledlow (Eds.), *Community preparedness and response to terrorism*, pp. 243–68. Westport, CT: Praeger.

Semple, E. (2009). Update your crisis communication plan with social media. *Strategic Communication Management, 13*(5), 7.

Sorensen, J. H. (2000). Hazard warning systems: A review of 20 years of progress. *Natural Hazards Review, 1*, 119–25.

Sorenson, J. H., & Sorenson, B. V. (2006). Community processes: Warning and evacuation. In H. Rodriguez, E. L. Quarantelli, & R. R. Dynes (Eds.), *Handbook of disaster research* (pp. 183–99). New York: Springer.

Steffen, J. (2013, December 15). Arapahoe High students look for hope with classmate in coma. *Denver Post*. http://www.denverpost.com/news/ci_24730610/arapahoe-high-students-look-hope-classmate-coma.

Stephens, K. K., & Malone, P. (2009). If the organization won't give us information...: The use of multiple new media for crisis technical translation and dialogue. *Journal of Public Relations Research, 21*(2), 229–39.

Taylor, M. (2010). Organizational use of new communication technology in product recall crises. *The handbook of crisis communication* (pp. 410–21). Malden, MA: Wiley-Blackwell.

Taylor, M., & Kent, M. L. (2007). Taxonomy of mediated crisis responses. *Public Relations Review, 33*(2), 140–46.

Thayer, K. (2013, November 9). Chicago-area Filipinos contact family affected by Typhoon Haiyan. Retrieved from http://www.chicagotribune.com/news/local/ct-met-philippines-typhoon-chicago-20131109,0,6991809.story.

Turoff, M., Chumer, M., Van de Walle, B., & Yao, X. (2004). The design of a dynamic emergency response management information system (DERMIS). *Journal of Information Technology Theory & Application, 5*(1), 1–35.

Veil, S. R., Buehner, T., & Palenchar, M. J. (2011). A work in process literature review: Incorporating social media in risk and crisis communication. *Journal of Contingencies and Crisis Management, 19*(2), 110–22.

Vieweg, S., Palen, L., Liu, S., Hughes, A., & Sutton, S. (2008). Collective intelligence in disaster: An examination of the phenomenon in the aftermath of the 2007 Virginia Tech shootings. *Proceedings of the Information Systems for Crisis Response and Management Conference*. Washington, DC.

Weick, K. E. (1995). *Sensemaking in organizations*. Thousand Oaks, CA: Sage.

Wright, D. K., & Hinson, M. D. (2009). Examining how public relations practitioners actually are using social media. *Public Relations Journal, 3*(3), 2–32.

Xiang, Z., & Gretzel, U. (2010). Role of social media in online travel information search. *Tourism Management, 31*(2), 179–88.

Yang, S., & Kang, M. (2009). Measuring blog engagement: Testing a four dimensional scale. *Public Relations Review, 35*(3), 323–24.

Yates, D., & Paquette, S. (2011). Emergency knowledge management and social media technologies: A case study of the 2010 Haitian earthquake. *International Journal of Information Management, 31*(1), 6–13.

Yeomans, M. (2012, November, 6). Social media's crucial role in disaster relief efforts. From www.theguardian.com/sustainable-business/social-media-hurricane-sandy-emergency-planners.

Yuan, Y., & Detlor, B. (2005). Intelligent mobile crisis response systems. *Communications of the ACM*, 48(2), 95–98.

Additional Resources:

- http://articles.washingtonpost.com/2012-11-20/national/35507983_1_social-media-staten-island-hurricane-sandy
- http://www.myfoxatlanta.com/story/24575661/woman-forms-facebook-group-to-help-others#axzz2s0acairM
- http://www.nbcnews.com/technology/facebook-helps-good-samaritan-rescue-pregnant-woman-freezing-atlanta-storm-2D12019300
- https://images.search.yahoo.com/search/images;_ylt=AwrTcXuHA.xSFOsAcVOJzbkF?ei=UTF-8&n=60&y=Search&fr=yfp-t600&p=snowed+out+atlanta+facebook+page&fr2=sp-qrw-corr-top&norw=1
  http://www.cnn.com/2014/01/30/us/atlanta-ice-how-it-happened/index.html

Possible Case

FEMA: Using social media to alert publics during Atlanta snowstorm

Source: https://www.fema.gov/news-release/2014/02/12/fema-continues-support-response-efforts-severe-winter-weather

WASHINGTON—The Federal Emergency Management Agency (FEMA) continues to closely coordinate with impacted and potentially impacted states in the path of a severe winter storm, through its National Response Coordination Center in Washington D.C., and its regional offices in Atlanta, Boston, New York City, and Philadelphia.

Today, President Obama declared an emergency for all counties in the state of South Carolina, at the request of Governor Nikki Haley, authorizing FEMA to support the state in its efforts to respond to the storm. The declaration comes in addition to the president's Emergency Declaration for 91 counties in the state of Georgia yesterday, at the request of Governor Nathan Deal.

FEMA has deployed an Incident Management Assistance Team to the Georgia Emergency Operations Center in Atlanta, along with liaisons to the state emergency operations centers in Georgia, Maryland, Pennsylvania, South Carolina, and Virginia to facilitate close coordination with the states. FEMA has activated its Regional Response Coordination centers in Atlanta and Philadelphia, and continues to be in close contact with state, tribal, and local partners in impacted and potentially impacted areas and stands ready to support its partners, if requested and needed.

FEMA has also established an Incident Support Base in Augusta, Georgia, where additional federal teams are on the ground. Commodities including generators, meals, water, blankets, and cots are being moved to that location. At all times, FEMA maintains commodities, including millions of liters of water, millions of meals, and hundreds of thousands of blankets strategically located at distribution centers throughout the United States and its territories, including Atlanta, Georgia, and Frederick, Maryland, if needed and requested.

The U.S. Department of Transportation's Federal Highway Administration is helping facilitate the expedited movement of utility trucks and personnel in Florida, Georgia, Mississippi, and South Carolina which includes bypassing weigh stations as long as they are under the legal weight requirements.

According to the National Weather Service, dangerous ice and snow are expected to intensify this evening as the storm moves up the Eastern Seaboard, affecting locations across the mid-Atlantic and Northeast. More than one inch of ice accumulation is possible from central Georgia into South Carolina through Thursday morning. Residents along the path of the storm can find their local forecast at www.weather.gov.

When natural disasters like severe weather strike, the first-responders are local emergency and public works personnel, volunteers, humanitarian organizations, and private organizations that provide emergency assistance required to protect the public's health and safety and to meet immediate human needs.

FEMA encourages residents and visitors in the track of the storms to follow the instructions of state, local, and tribal officials, and monitor NOAA Weather Radio and their local news for updates and directions provided by local officials. Residents can find trusted sources for weather and preparedness information via Twitter on FEMA's Social Hub at http://www.fema.gov/social-hub

Wireless Emergency Alerts are currently being sent directly to many cell phones on participating wireless carrier networks. These alerts are sent by public safety officials such as the National Weather Service about imminent threats like severe weather. They look like a text message and show the type and time of the alert, any action you should take, and the agency issuing the alert. More information on Wireless Emergency Alerts is available at http://www.ready.gov/alerts. Individuals can check with their cellular carrier to determine if their phone or wireless device is WEA-enabled.

FEMA encourages all Americans to visit www.ready.gov or www.listo.gov to learn more about how to be better prepared and how to protect your family during emergencies.

Generator Safety

Carbon monoxide or CO is a colorless and odorless gas that is emitted from fuel-burning appliances, like generators, or machines that are not working or venting properly. Breathing in high levels of carbon monoxide can be fatal and kills more than 150 Americans annually. FEMA recommends the following steps to protect your family from the dangers of carbon monoxide:

- Install and maintain CO alarms inside your home to provide early warnings
- Install CO alarms in a central location outside each separate sleeping area and on every level of your home
- Use portable generators outdoors in well-ventilated areas away from all doors, windows, and vents
- Make sure vents for the dryer, furnace, stove, and fireplace are clear of snow and other debris, and
- Remove vehicles from the garage immediately after starting.

For more information and winter preparedness tips, please visit www.usfa.fema.gov to find out more on carbon monoxide and fire safety.

Preparing for Severe Winter Weather

Get to know the terms that are used to identify winter storm hazards and discuss with your family what to do if a winter storm watch or warning is issued.

- A **Winter Weather Advisory** means cold, ice, and snow are expected.
- A **Winter Storm Watch** means severe weather such as heavy snow or ice is possible in the next day or two.

- A **Winter Storm Warning** means severe winter conditions have begun or will begin very soon.
- An **Ice Storm Warning** is when freezing rain produces a significant and possibly damaging accumulation of ice.
- **Freezing Rain** creates a coating of ice on roads and walkways.
- **Sleet** is rain that turns to ice pellets before reaching the ground. Sleet also causes roads to freeze and become slippery.

Avoid traveling by car, but if you must, make sure you have an emergency supply kit in the trunk of your car. FEMA encourages families to maintain an emergency supply kit both at home and in the car to help prepare for winter power outages and icy or impassable roads.

An emergency supply kit should include a three-day supply of food and water, a battery-powered or hand-crank radio, and extra flashlights and batteries. Thoroughly check and update your family's emergency supply kit and add the following supplies in preparation for winter weather:

- Rock salt to melt ice on walkways;
- Sand to improve traction;
- Snow shovels and other snow removal equipment; and
- Adequate clothing and blankets to help keep you warm.

Ensure your family preparedness plan and contacts are up-to-date. Learn about the emergency plans that have been established in your area by your state and local government, and ensure your home and car are prepared for the winter weather.

For more information and winter preparedness tips, please visit www.ready.gov/winter-weather or www.listo.gov to find out how you can prepare your family for winter storms and other disasters.

FEMA's mission is to support our citizens and first-responders to ensure that as a nation we work together to build, sustain, and improve our capability to prepare for, protect against, respond to, recover from, and mitigate all hazards.

Follow FEMA online at www.fema.gov/blog, www.twitter.com/fema, www.facebook.com/fema, and www.youtube.com/fema. Also, follow Administrator Craig Fugate's activities at www.twitter.com/craigatfema.

The social media links provided are for reference only. FEMA does not endorse any non-government websites, companies, or applications.

Discussion Questions:

1. How can the city of Atlanta use social media to address the affected population?
2. How can the city of Atlanta and its emergency responders use social media to glean information from residents that can be used in the crisis management effort?
3. Assess or discuss the adequacy/inadequacy of the FEMA alert system for the Atlanta snowstorm.

· 5 ·

RESEARCH IMPLICATIONS FOR CCCS

Chapter Highlights

- Relevant research on community-based crisis communication and management
- Implications for future research

Establishing crisis communication centers in communities is a prudent step in being crisis-ready. However, the justification and commitment for such a center is often difficult to find, especially when dealing with limited and scarce resources. Notwithstanding, the ideas and principles of anticipatory models of crisis management (AMCM) suggest that the price of not planning for crises and disasters is often greater than the amount of resources it will take to put in place adequate planning. Hence, this chapter attempts to provide a further justification for establishing a CCC, but, more importantly, it looks at the research implications of establishing such a center.

A community contemplating the establishment of a CCC must first identify and address what constitutes a crisis and what doesn't. The community of scholars and researchers can help in this area by the nature of the crisis they choose to study or to pay attention to. At the same time, it needs to be said that certain issues facing traditional and for-profit organizations do not necessarily apply to CCCs, because the aim is not so much focused on image reparation per se, but on addressing human issues or restoring a community to normal activity and engagement.

From this standpoint, the use of the terms *crisis*, *disaster*, and *emergency* independently or interchangeably in the literature, muddles how we view crises and what needs to be done when they occur. For instance, natural disasters are often considered an act of nature and are thus not preventable, which then impacts how people respond to them. At the same time, some people hold the perspective that all forms of natural disasters are caused by humans, indicating some degree of preventability (see Dorasamy, Raman, & Kaliannan, 2013; Shrivastava, Mitroff, & Alpaslan, 2013). Furthermore, Palen, Vieweg, Liu, and Hughes (2009) contend that certain events and, in particular, disasters, such as wildfire and hurricanes, are concrete challenges faced by communities and by the public, which leaves no doubt as to their characterization as crises, but these events also put the public and responders in a precarious situation, as the specifics of collective intelligence called for are harder to pinpoint. Specifically, Palen, Vieweg, Liu, and Hughes (2009) argued that disasters, for instance, necessitate certain types of responses within organizations that do not necessarily happen in other kinds of emergencies. Therefore, they argued that the term *disaster* is typically not used to describe school shootings. Similarly, Perry and Quarantelli (2005) pointed out that the definition and usage of the terms *disaster* and *emergency* are not precise. Consequently, Palen, Vieweg, Liu, and Hughes (2007) suggested that *crisis* functions as an umbrella term that describes or consists of large-scale emergencies, disasters, and unexpected events. Quarantelli (2005) asked us to think about where terrorism fits into the conceptualization of a crisis. Is it a disaster or is it a catastrophe? He reiterated that some may look at terrorism as having distinctive characteristics, but others see it as a catastrophe.

Heath (2012) also addresses the impreciseness of the term *crisis* by stating that there are many definition of crises that have surfaced over the years. For example, Heath and Millar (2004) listed about 20 definitions. Similarly, Shrivastava, Mitroff, and Alpaslan (2013) also discussed the problem with how a crisis is conceptualized. Some of the crisis definitions focused on certain turning points in the life of organizations and others focused on the management effort to respond beyond the normal scope of their routine procedures. Other definitions allude to the level of stress accompanying crises, while still others deal with insufficient controls, uncertainties, and violations of laws or ethics, among other things.

Researchers have indicated that organizations are defined in terms of how they plan or respond to crises (Heath, 2012; Olaniran & Williams, 2001, 2012). In keeping with the principles outlined in the AMCM, planning for a crisis through anticipation is essential to the eventual crisis management. The ability to anticipate ultimately determines the effectiveness of a given crisis event and how the associated organization is judged. To this end, crises can be viewed as risk manifestation. Therefore, organizations in crisis are evaluated by the attention they paid to the risk management. It is true that not all crises are preventable; however, the anticipatory model suggests that organizations still have the responsibility of attempting to predict events that can contribute to or create a full-blown crisis. Putting it succinctly, Heath (2012) argued that "if it [organization] cannot prevent a crisis, [it can] at least understand the conditions and preparation requirements sufficiently to be prepared to respond, bring control, mitigate damage, and protect other interests" (p. 3).

Perhaps one way of overcoming the challenge with crisis conceptualization is to first recognize that regardless of terminologies, crises and disasters need not be seen as events but as processes. The AMCM presented in this book attempts to assess the nature of crisis by looking at the pre-crisis, crisis, and post-crisis stages, which are not necessarily independent of one another but are, rather, informed by and intertwined with one another. On the same note, it needs to be emphasized that the crisis process is characterized by preconditions that cause low-impact failures and then proceed to a crisis-triggering event that creates significant damages, which then necessitate, or are followed by, an emergency management or crisis response that culminates in rehabilitation and compensation of stakeholders (see also Jaques, 2007, 2009; Olaniran & Williams, 2001, 2012; Seeger, Sellnow, & Ulmer, 2003). More important is the fact that failure to do a good job in any of the processes would result in crisis management being perceived as ineffective (Olaniran & Scholl, 2014; Olaniran, Scholl, Williams, & Boyer, 2012; Shrivastava, Mitroff, & Alpaslan, 2013).

Just as it is true for organizations, it is the same for communities. A community has the burden of responsibility to protect itself and/or citizens from crises. Communities need to attempt to foresee its areas of vulnerability and crisis susceptibility and then mobilize its citizens or crisis responders to put in place mechanisms for addressing the potential crisis. In other words, a crisis should be planned for and controlled, and that is the major task that a CCC should undertake, and how the center is assessed.

Control

One of the key components of the AMCM is control. Control, or the degree to which a CCC organization has a handle on a particular or potential crisis, is an essential part of the process. It is not enough to wait for a crisis to occur before a CCC puts in place activity or measures to address the crisis. Rather, it is germane that the CCC views and accepts its obligation to understand that crisis management is an ongoing task. Furthermore, while one crisis is being dealt with, there is always the potential that another one may be brewing, and that could then impact how crisis-ready a CCC and its community members are. The fact that available literature almost always alerts the public to a particular incident is detrimental to the discipline of crisis management (Heath, 2012; Shrivastava, Mitroff, & Alpaslan, 2013). For instance, focusing on a particular event forces or conditions us to think of incident-specific crises, which in essence prevents us from looking at a crisis as more than an event. As a result of this oversight, we then tend to miss the bigger picture and other factors that underlie crises in general. For example, there are crises that do not manifest themselves in terms of deaths or injuries, but rather in terms of the general wellness of a community. For instance, an adverse decision or mistake made by a single organization within a community may create a crisis. For instance, when Cargill Meat Solutions, the largest employer in the town of Plainview, Texas, closed its plant, it had ripple effects in the town of 22,230—its workers had to look for a new source of employment. As part of the effect of the plant closure, another company (i.e., Packers Sanitation Services, Inc.,), which cleaned the beef packing plant had to lay off 125 workers when Cargill closed its operations (Welch, 2013). When Cargill closed its plant, more than 10% (i.e., 2,300) of the entire population lost their jobs (Diaz, 2013), which created a major crisis for the town in terms of how to find placement for workers so that the city did not turn into a ghost town. Not only does the city have to address the issue of finding placement for the displaced workers, but city leaders also have to deal with the question of what will happen when unemployment benefits run out (Diaz, 2013), and adjust to the reality of dwindling tax revenues while trying to keep the city afloat. According to one estimate, the Cargill plant closure was believed to have cost about $1.1 billion to the region's economy (Welch, 2013). However, beyond financial impact, there are human issues, such as family displacement and survival that cannot be measured in dollar terms. More importantly, in these kinds of crisis scenarios we are reminded that it is not whether there is a crisis, but rather which or what crisis is deserving of our attention (Heath, 2012). To this end, organizations and communities alike face on a day-to-day basis the need to identify, to plan, and to control crises and potential crises by having to prioritize what issue deserves their attention. At the same time, one needs to understand that even when an organization such as the CCC has in place a crisis plan or is perceived to be in crisis, its plan may not be applicable to the actual crisis (Jaques, 2009; Shrivastava, Mitroff, & Alpaslan, 2013). In order words, the AMCM notion of control would suggest that an organization and, in particular, the CCC, needs to have the mind-set that with crisis and disaster management crisis plans should be updated frequently with new information as it become available. More specifically, it would not hurt to take a different position from the "if it ain't broke, don't fix it" mode to more of a "if it ain't broke, fix it anyway" approach, because one cannot be 100% prepared for crises or disasters.

There appears to be no easy or cookie-cutter solution that will help all communities in terms of their crisis preparedness or crisis and disaster management. However, precedents have been set by organizations, such as the American Red Cross and Samaritan Purse, charged with coordinating people and their activities to handle crises and disasters regardless of geographic location. CCCs can learn from organizations like the Red Cross, realizing that it coordinates volunteers by flying them into disaster locations to address a series of unspecialized and nonspecific tasks—including managing and mobilizing their volunteers (Palen, Vieweg, Liu, & Hughes, 2009). The point is that there will be activities for CCCs that include several rescue and recovery tasks to be done for any given crisis. Notwithstanding, the CCC's job is not to be an expert on all issues but to serve as a conduit, albeit remotely, with the responsibility for coordinating and disseminating information to individuals and to other organizations that are helping or that are responsible for addressing a particular crisis and disaster. Palen, Vieweg, Liu, and Hughes (2009) suggest that the production and dissemination of information and tasks can happen remotely in a centralized and decentralized manner as a way to leverage a large pool of planned and emerging volunteers. For example, in previous chapters we stressed the importance of a CCC serving as a command and control center that deals with the logistics of identifying shelters and occupancies, determining medical and personnel resources, and communicating that information to the publics and key officials in an attempt to help mitigate crisis impacts. Similarly, part of the tasks for the CCC is to implement or to take advantage of technology, including the use of databases or what has been termed as an *emergency management information system* (EMIS), to coordinate and integrate emergency responders, workers, and volunteers (Dorasamy, Ramon, & Kaliannan, 2013). It is also important for a CCC to learn how other communities handle their crises as a way to update its crisis preparedness and plan.

The recognition of a CCC as a formal and authoritative institution responsible not only for information production but also for dissemination is called for. The shift from an information-dissemination role to that of information collection, and the simultaneous attention to both of these roles, needs to be emphasized and addressed in the literature. After all, the landscape of crisis and disaster management has changed and is changing from the exclusive use of traditional media to the one where incorporation and adoption of new communication media are imperative. Thus, researchers can, and ought to, focus on the changes created by the shift in perspective. For instance, crisis and disaster management in general, and specifically with the CCC, should highlight and discuss how key officials, and the center as a whole, include or utilize the *distributed network*, especially with regard to information sources and collective skills produced by the public. It has been suggested that analysis of the information distribution and network communication can be subjected to crisis informatics research (Dorasamy, Raman, & Kaliannan, 2013; Palen, Vieweg, Liu, & Hughes, 2009). There is no doubt that crisis informatics can illuminate our understanding of the vast amount of data available to the CCC and to other crisis practitioners. Yet, other noninformatics researchers can elucidate the type of communication activity and its relationship within a vast array of communication networks available to the center and to publics. In essence, this would help foster the sensemaking process about a crisis or disaster management while helping to identify a reliable and coherent theme from the available data (Dorasamy, Raman, & Kaliannan, 2013; Lukaszewski, 2013).

Methods

From the methodological standpoint, the literature on crisis management is rife with case studies or specific crisis events (An & Cheng, 2010; Cutler, 2004; Olaniran & Scholl, 2014; Shrivastava, Mitroff, & Alpaslan, 2013).

From a social science standpoint, case studies are useful but challenging. For instance, case studies allow a researcher or community of scholars to gather information about a particular crisis event, for instance, individuals, organizations, and the accompanying product or service (Distaso & Stacks, 2010; Stacks, 2006). Case studies, however, may be inadequate in terms of capturing the multiplying effect of a particular crisis on another crisis. For example, it is imperative that we not only look at a crisis in terms of what an organization did wrong, but also how what an organization did well, either of which may create a crisis for other organizations (Heath, 2012; Shrivastava, Mitroff, & Alpaslan, 2013). Heath (2012) illustrated with the example of the Sarbanes–Oxley legislation passed by Congress to avoid the kind of financial crisis created by the likes of Enron and WorldCom by requiring publicly traded companies to follow more stringent financial reporting.

While incident specifics and the use of case studies are not going to go away in crisis and disaster management, it appears prudent that researchers engage in other methodologies that can overcome the challenges with case studies. For example, Distaso and Stacks (2010) recommend the use of participant-observer technique. The participant-observer method is less used in crisis management but attempts to focus on the day-to-day activities of corporate members, which allows researchers to gain insights into how they think and how they behave. Shrivastava, Mitroff, and Alpaslan (2013) also allude to the use of role-playing, especially in teaching students or would-be crisis practitioners about moral and ethical implications that crisis and disaster management may entail for self-assessment and stakeholders. More importantly, such a role-play initiative can help teachers help students realize the importance of crisis prevention by distinguishing between it and crisis management. In essence, having such an opportunity could provide adequate insights into crisis precursors or preparation plans. The participant-observer method could shed some light on potential problems facing an organization and their clients, and the interaction of corporate officials with their respective publics (Michaelson & Stacks, 2010). This is a scenario that may help students understand how to resolve conflicts resulting from different viewpoints by multiple stakeholders and understand that crises affect multiple stakeholders (Shrivastava, Mitroff, & Alpaslan, 2013). Such an experience would also provide students with invaluable lessons about issues of crisis communication and decision making (Shrivastava, Mitroff, & Alpaslan, 2013; Wallace & Weber, 2010). Specifically, the participant-observer method would benefit the CCC by allowing crisis practitioners the opportunity to observe and to assess how CCC staffs handle their daily responsibilities at each stage of crisis disaster management. Also, when appropriate, recommendations can be made to improve or to correct mistakes and challenges faced by these centers.

Palen, Vieweg, Liu, and Hughes (2009) expressed similar methodological concern about the use of qualitative data in crisis management. In their study of Facebook use, the researchers expressed the view that manual qualitative data collection and analysis methods only allow the ability to capture a moment in time or a *brief event*, even when the event is of a national importance. Hence, Palen, Vieweg, Liu, and Hughes (2009) suggested that such an approach "is not a scalable or sustainable approach for ongoing sociological investigation of emergency response" (p. 477). Consequently, there is a need to incorporate other tools being developed by the community of scholars in the e-social science community for cyber-enabled data collection and analysis. Furthermore, given that crisis and disaster events transcend geographic, temporal, and virtual issues, they therefore necessitate newer instruments and/or tools that are more powerful data-gathering and analytical solutions from core crisis informatics research (Dorasamy, Raman, & Kaliannan, 2013; Palen, Vieweg, Liu, & Hughes, 2009). In essence, data collection and analysis tools need to take advantage of new multimodal communication and information technologies. For instance, tools that enable the integration of multimodal data sources will include not only media-provided data but also things like an eyewitness video clip, information from multiple social media sites, and alternative news updates, along with official notifications in a consistent and coherent manner. Consequently, replayable information, such as the Digital Replay System (DRS) (French et al., 2006; Greenhalgh, French, Tennent, Humble, & Crabtree, 2007) can be helpful. These technological capabilities are believed to offer researchers who are geographically disadvantaged from crisis and disaster sites the ability to overcome such disadvantages, specifically, the fact that crises and disasters are geographically bound but the participation of individuals in such events can flourish (i.e., digital participation). Thus, researchers or digital participants can gather data from the crisis sites using available streaming video or what is called *environmental sensor data* (Palen, Vieweg, Liu, & Hughes, 2009).

Technology and New Media Use in CCCs

The explosion of new communication technologies facilitated by mobile devices and social communication networks holds critical implications for crisis and disaster management. We are only at the beginning stage of understanding the roles that these new contemporary communication technologies can play in crisis management. Therefore, past research can inform us on new media incorporation across crisis management stages. Not only that, past research on the use of new media can inform crisis management practitioners on the new directions that require our attention now and in the future. For instance, research indicated that crisis communication messages are more effective when they have a personal touch and offer action steps (Dillard, Shen, & Vail, 2007; Heath, Lee, & Ni, 2009; Palenchar & Freberg, 2012). However, not a whole lot is known about the specifics of how to bring about such a personal touch. Thus, future research can take on this initiative by identifying how different media platforms can help to create a personal touch and a sense of immediacy, especially when one accounts for different communication media and how they support social cues, such as text, audio, and video. Understanding the new media effects and impacts in communication can then allow key crisis practitioners (i.e., CCC officials) a way to design risk and crisis messages in a manner that helps members of the public access the right message and perhaps increase message compliance.

Similarly, understanding the level of immediacy supported by different communication platforms may be instrumental in helping to nurture relationships between the CCC and its partner organizations. For example, in the crisis preparation and planning phase, issues such as how different partner organizations prefer to receive messages from the CCC and the required frequency of such communication can be negotiated prior to crisis or disaster onset. Recognizing the importance of new media in crisis management and public relations, Palenchar and Freberg (2012) provide a review of different areas of focus by researchers. For example, few researchers explored privacy and impression management (e.g., Hinduja & Patchin, 2008; Kramer & Winter, 2008; Walther, 2007; Walther, Van Der Heide, Kim, Westerman, & Tong, 2008). Other researchers have focused their attention on social networking profiles (e.g., Boyd & Ellison, 2008; Kramer & Winter, 2008; Zhao, Grasmuck, & Martin, 2008), while some researchers concentrate on how people construct and negotiate online identities using social media (Mazer, Murphy, & Simonds, 2007; Zhao, Grasmuck, & Martin, 2008). Another set of researchers has focused on the use of particular social media platforms, such as Facebook and Myspace. A case in point is the analysis of Facebook in the Virginia Tech shooting that combed through qualitative data and analyzed information directly linked to a national emergency (Palen, Vieweg, Liu, & Hughes, 2009).

Beyond the selection or deployment of appropriate communication technologies in crisis and disaster management, there is a pressing need to understand the underlying communication goals. While deploying new media in crisis communication may allow for more equally distributed information to members of the public or stakeholders, the adoption of these communication technologies, in themselves, holds potential for crises or hinders effective crisis management. For instance, the technical capabilities of new media inadvertently create overreliance on technologies, which clouds crisis managers' judgment regarding the fact that technologies do fail (Herbane, 2013; Olaniran & Williams, 2001, 2012) and prevents them from thinking about such risks (Palenchar & Freberg, 2012; Traynor, 2008). There are other inherent problems with these technologies, including system overloads and message delivery, or lack thereof, between crisis responders and the public or other organization partners, such as EMS systems. Furthermore, technology brings up the risk of inadvertently releasing sensitive information not meant for public consumption. All these point to the fact that emergency management professionals may be inadequately prepared or trained in the implementation of new media in crisis communication plans (Nichols, 2010). Therefore, there is a need for organizations in general, and, more specifically CCCs, to provide adequate training for staff and if possible its partnering organizations. This is essential because it has been established that most crisis and risk communication training emphasizes corporate briefings and spokespersons' dos and don'ts without specific emphasis on technologies (Palenchar & Freberg, 2012). Failure to train workers and crisis management professionals may result in troubling consequences, such as overloading the system's bandwidth, inadvertently crashing the communication network, and disrupting the knowledge management systems (Dorasamy, Ramon, & Kaliannan, 2013; Stephenson & Bonabeau, 2007). Furthermore, there is a need for training crisis professionals in how to handle or to manage multiple communication media and deciding on the types of communication appropriate to be disseminated through them, which consequently would require proper planning and preparation (Palenchar & Freberg, 2012).

Other areas of technology worthy of attention are cyberattack and security. The continued dependency on computers and technologies subjects not only protects nations but also communities and their infrastructure (e.g., bioterror and hacking of aviation procedures) from foreign and domestic organizations or individuals. Unfortunately, most of the attention and initiatives to address cybersecurity are at either the national or private organizational levels. Communities however, need to do a better job of preparing, countering, responding, and recovering from cyberattacks as their stakeholders would be the ones affected when such attacks occur. For example, the Internet of things (IOT) highlights the cybersecurity danger because hackers are able to hijack vehicles remotely, which could create log jams and accidents on community highways. The dangers may not be so much in accessing a particular device or object, but rather in the irreparable damages that hackers may be able to unleash once they find an entry point within a network (Brodbeck, 2015). It is, therefore, imperative that the CCC helps communities in involving public and private sectors and key experts in training and communicating the risk of using computer networks and network safety. CCCs, along with academic experts, can help design training on cybersecurity that will focus on reinforcing the role of communities, private sectors, and individuals in recognizing the risks and protecting against cyberattacks (Shields, 2015; Zhao & White, 2014).

Research implications include figuring out how organizations conduct the necessary training and/or providing organizations with the best practices for providing such training. This should, however, move beyond mere theoretical prescription to application. In essence, the CCC and other organizations dealing with crisis and disaster management must work together to figure out how to implement effective technological training. An area worthy of consideration and focus by researchers is how to implement technology use while guiding against information overload. Another area of focus might be how to navigate a multitude of social media channels without losing the focus on communication goals. This is an important challenge, especially because the social media channels by nature are inherently risky. For instance, with social media and SNS, anybody can post anything online regardless of whether they are true or not. Consequently, the CCC must figure out a way to maintain a delicate balance by providing necessary information while ensuring the credibility of such information. Establishing such credibility would allow or reduce publics' need to look for alternate sources of information and consequently limit the attention paid to rumors and inaccurate information. In other words, organizations such as the CCC must communicate in a transparent manner and understand its role and the challenges it faces in the digital age and community (Heitmann & Lott, 2008; Waters, Burnett, Lamm, & Lucas, 2009).

The Interdependent Relationship in Crisis Stages

Crises, disasters, and emergencies, as presented in this book, can be examined from the standpoint of the three stages of crisis management, which include pre-crisis, crisis response, and post-crisis stages. Traditionally, the majority of organizational effort has been dominated by a bias focused on crisis response. However, the three stages of crises are not mutually exclusive propositions but rather are interdependent components in any crisis or disaster management for effectiveness. We concur with Heath (2012) that understanding the interdependency of these stages can advance the field of crisis management through theory development, research, and professional practices. First of all, the pre-crisis phase concentrates on crisis planning and preparedness by trying to answer the question of what can be done to reduce the likelihood of a crisis or at least reduce the effects of a crisis when it occurs (Coombs, 2009, 2014; Heath, 2012; Olaniran & Williams, 2001, Scholl, Williams, & Olaniran, 2005). Specifically, this implies that a high-risk organization conducts an assessment of its operational strategies and communicates with its environment or public the potential risks that can result, and stresses its plan to prevent and to alleviate damages. In a similar manner, an organization such as the CCC that is not a corporate entity but operates as a layer of community safety mechanism must assess risks emanating from corporate entities and natural disasters and then provide members of its community with critical information for averting a crisis or reducing the impact of a crisis. Heath (2012) calls this a foundation for effective community relations and effective crisis communication. Olaniran and Williams's (2001) anticipatory model of crisis management is predicated on effective crisis or pre-crisis planning, while Coombs (2009) also suggests that pre-crisis should focus on prevention and preparation.

Critical to any crisis management, however, are other variables including threats, risks, and fear (Jin, Pang, & Cameron, 2012). Threat is characterized by a warning to recognize danger and acceptance of suggested recommendations to avoid negative consequences (Jin et al., 2012; Strong, Anderson, & Dubas, 1993). Fear, on the other hand, is considered as an outcome of a *perceived threat* (Jin et al., 2012). Risk is not the actual loss but rather expected loss. Risk is consequentially inherent in risk communication messages by decision makers. As a result, gathering and processing information about a crisis or a potential crisis is helpful in reducing uncertainties associated with such a crisis (Oppe, 1988). Unfortunately, the challenge is that a real loss is only ascertained after a crisis has happened (Jin et al., 2012). The uncertainty associated with not knowing the real loss has been linked to failure to put a crisis management action plan in place, especially when considering the cost of such a plan for an event that may or may not occur (Herbane, 2013). However, it has been suggested that the cost of a crisis is almost always and substantially more than the cost of crisis planning (e.g., Olaniran & Williams, 2001, Scholl, Williams, & Olaniran, 2005). From a different standpoint, the more severe a threat or perceived threat is, the greater is the motivation to embrace avoidance. At the same time, the more a perceived vulnerability to a given threat, the more the likelihood of compliance (Jin et al., 2012; Palenchar & Freberg, 2012).

A major question facing decision makers in crisis management is how do these variables (i.e., threats, risks, and fear) play out in the crisis management stages? The answer is apparent in that these variables play key roles across all the three stages of crisis management. For instance, there's the fact that the pre-crisis stage focuses on planning and preparedness; it then stands to reason that information regarding threats and risks must be communicated from organizations to its environment or to the public. All in all, communities like organizations must realize that they are not isolated from crises or disasters that seem far from them (Herbane, 2013). For example, the Ebola episode of 2014 seemed not at all transferable to the United States and to European countries in its early stages, but ultimately the epidemic spread to these nations and thus brought the realization that Ebola is not just something that affects some countries in West Africa. Thus, communities stand a greater risk of crisis from increased transborder events (Herbane, 2013; Smith & Fischbacher, 2009) that may escalate if adequate crisis planning cannot be put in place through community-based CCCs. Hence, communication focus needs to be on risk messages, otherwise known as risk communication.

On the other hand, the crisis response stage is not devoid of risk, and communication to stakeholders and to the public must accurately change from the speculative to the specifics of the reality. Hence, publics demand and crave helpful information (e.g., an action plan) that can allow them to cope with the occurring crisis. The CCC would then emphasize the importance of compliance with crisis messages to cope with a given crisis while mediating damages. For example, during a recent polar vortex, which brought with it extreme cold-weather storms for several days, the messages from national and local weather forecasters stressed the importance of staying home and avoiding travel unless absolutely necessary. The risk of traffic accidents and power outages was conveyed to the affected publics and created a fear of the results of noncompliance, such as fatalities from accidents. The post-crisis phase of crisis management centers on lessons learned from a given crisis or disaster. The post-crisis phase provides an opportunity to assess the effectiveness of communication strategies in the prior two stages. The post-crisis phase can give voice to various individuals, including management and the public (Heath, 2012), but primarily they should reinforce communication messages regarding existing threats that still prevail even after a crisis episode. Fear and risk of the crisis reoccurring, or a different type of crisis altogether occurring, should be presented to the community while providing action plans for avoiding any potential crises that may be forthcoming. It is important, however, for the CCC to recognize the ongoing nature of its task so that the end of a crisis merely ushers in the beginning of the crisis management process all over again, taking into consideration both internal and external factors that can serve as crisis-triggering mechanisms (Olaniran & Williams, 2012; Shrivastava, Mitroff, & Alpaslan, 2013).

When one places a crisis in the context of the AMCM and the stages of crisis and disaster management, it is imperative that one understands that the CCC's role is one in which risk and threats are managed. Consequently, the AMCM provides a platform for doing that. Issues or issues management is believed to emanate from risks. More specific, however, is the idea that issues management is *contestable* and inherent in value, policy, and identification. Issues management is therefore related to policy and agenda setting in the process of crisis management or in avoiding crises altogether (Jaques, 2012; Shrivastava, Mitroff, & Alpaslan, 2013). An issue can become a crisis. An example is the Lubbock community, where synthetic marijuana that was supposed to be an alternative to illegal marijuana created a cause for concern for local authorities and for parents, who claimed that the product damages the brain and, consequently, were able to galvanize protest against tobacco shops, leading to eventual banning of the synthetic drug in the community. Similarly, while there are no official words from the medical community regarding e-cigarettes, the same challenge that cigarette companies faced may await electronic cigarette manufacturers in debates regarding the unknown health hazards of e-cigarettes use.

Expectation and Control

At the center of any crisis management is the issue of expectation and control (Olaniran & Williams, 2001). The idea of what stakeholders expect or how they appear to hold an organization accountable is germane to any crisis and is what the CCC must deal with. Expectations ostensibly guide what is communicated and what organizations should do to mitigate crises. Stakeholders tend to hold whoever is responsible for a crisis to these standards (Lukaszewski, 2013; Shrivastava, Mitroff, & Alpaslan, 2013). Hence, the CCC is advised to do its due diligence in meeting and in exceeding reasonable public expectations to keep the community and its citizens safe. Similarly, expectation is related to control, which is the degree of influence a crisis organization exerts over a crisis or in preventing one from happening. Control again is measured in terms of public expectation and their need to know whether an organization has done everything in its power to avoid crises beyond mere image restoration plans. To this end, some have called for crisis preparedness and, in particular, the crisis management plan (CMP) as an essential tool in gaining control of crisis and/or disaster management (e.g., Coombs, 2000, 2014; Low, Chung, & Pang, 2012; Herbane, 2013; Olaniran & Williams, 2012; Seeger, Sellnow, & Ulmer, 2003, 2012).

It is imperative that a crisis management plan be put in place, especially if one subscribes to the notion that crises are inevitable, but one is never sure of when they may occur. Thus, organizations, including the CCC, need to plan for the possibility of crisis occurrence. Strategic planning that includes anticipation of crises and their conditions, along with identifying key resources, structures, and strategies needed to resolve the crisis, is essential (Herbane, 2013; Low, Chung, & Pang, 2012, Olaniran & Williams, 2001, 2012; Pang, Cropp, & Cameron, 2006; Seeger, Sellnow, & Ulmer, 2003). The reality, however, is that fewer organizations have a crisis management plan than should (Burnett, 1998; Cloudman & Hallahan, 2006; Fink, 1986). An organization's measure of crisis preparedness is attributed to its crisis management plan. After all, the goal of crisis planning is to reduce, if not totally remove, uncertainties surrounding decision making about crises in terms of time, response, and recovery process (Leibinger, 1997, Seeger, Mitroff, & Alpaslan, 2003). Crisis plans demonstrate the importance of communication that allows or prepares crisis managers and public relations practitioners to be ready in addressing stakeholders' questions regarding what happened, how it happened, and how the crisis might affect them (Marra, 1998). In other words, in the face of a crisis, crisis management plans present a useful tool that can help effectively facilitate a competent response strategy (Gonzalez-Herrero & Pratt, 1996; Low, Chung, & Pang, 2012; McConnell & Drennan, 2006). A crisis management plan, however, needs to be actionable; it cannot be a document that is put in place merely for the sake of having a document. Some have emphasized that such document needs to be brief and user-friendly (Coombs, 2014; McConnell & Drennan, 2006). We believe that a good crisis management plan needs to address specific types of crises as a priority, but it may also incorporate guidelines specific to any crisis.

With control, the idea of issues management is essential in the sense that even when a particular crisis has been resolved or no longer seems to linger in the media, specific issues relating to the crisis may still linger (Jaques, 2012). For example, coverage of the school shootings, in Columbine, Newtown (Connecticut), and Arapaho High School appears to have dwindled in the media. However, for gun manufacturers or for industries or communities that pride themselves in their ability to own guns, either for personal safety or for hunting, the uncertainty and crisis potential increases regarding the possibility of what the negative attention to guns will do to their right to own guns. Therefore, the industry and rural communities affected by such issues must continually work to gain control of the issue in the media, just as much as the antigun folks are trying to get gun regulations passed. Accordingly, Heath (2010) suggests that in attempting to gain control, a crisis organization, regardless of whether it faces reputational or issues debates, needs to control the messages surrounding the issues involved. To this end, control is an outcome variable; however, one should not forget that control is also a process to reach a desired or specified outcome. Hence, control is multifaceted because it includes a way to address how much institutions and decision makers are in command of crisis situations. Control must also be observed or measured in relative terms, which include relevant environments, and stakeholder and public perceptions of a crisis or disaster event. More specifically, control is necessitated and called for in communication interactions, messages, and strategic actions designed to prepare and to implement effective crisis management. This can be achieved through information dissemination and media relations efforts with built-in flexibility to make changes whenever necessary (i.e., rigidity avoidance).

Given the focus on the CCC, it is important to recognize that it is not sufficient to examine only how crisis organizations seek control, but also how communities seek and negotiate control in crisis and disaster management. This major shift in focus is what might help a community recover from crises, disasters, and or emergencies quickly. The process has been argued as a way to bring "certainty to uncertainty or order to disorder" (Heath, 2010). In essence, it is the responsibility of a CCC to help the community and its citizens navigate conflicting messages emanating from a particular crisis and also help publics shape their understanding of the events, which will ultimately help determine the effectiveness of the crisis management. From this standpoint, we can increase our understanding of factors relating to issues management, risk, communication, messages, and shared control in crisis scenarios, among other important factors, all which impact effective crisis management (Coombs, 2014; Dorasamy, Ramsan, & Kaliannan, 2013; Heath, 2004, 2010; Shrivastava & Statler, 2012). As we grapple with the idea of control in both crisis and disaster management, we are reminded that control is a complex idea just as much as the discipline of crisis management itself. As a matter of fact, we are reminded by scholars of the need to have a paradigmatic shift where scholars and researchers move away from linear thinking to a more complex examination of crises, because it is a better way to have a handle on the complex nature of crises and to maintain the necessary adaptation and improvisation required to have control (Gilpin & Murphy, 2008; Heath, 2010; Herbane, 2013).

As far as control is concerned, one of the major takeaways for practitioners in crisis and disaster management is that issues and crises are seldom resolved in the short term (Jacques, 2010). Therefore, management and scholars alike should focus not only on a quick fix but also on a long-term impact of crises for organizations, communities, and their publics or stakeholders. The longer-term orientation would then help with assessment of feedback and learning, and consequently enable organizations and/or communities to be more adept at preventing and coping with future crises.

Bringing It Together

For the CCC to effectively deliberate its tasks or duties, there needs to be a recognition that it is an entity that operates both independently an interdependently with other organizations in a given community. While the CCC is charged with risk assessment, crisis planning, preparedness, response, and post-crisis management, the extent to which it can effectively handle each of these tasks is dependent on how well it is able to collaborate and work with government agencies (i.e., city and national) and other nongovernment organizations (NGOs) within the community. In other words, *community relations* efforts are the glue that can hold all the different tasks together. By community relations efforts, we imply the public relations effort where the CCC reaches out to manage the existing communication interaction networks while developing new communication interaction networks among organizations and publics for the sole purpose of creating win-win relationships (Neff, 2005). Community relations efforts have been considered the centerpiece in addressing crisis management, prevention, mitigation, and response (Coombs, 2014; Heath & Coombs, 2006; Heath & Ni, 2010). Similarly, others have identified the interdependency of organizations and their communities in community relations and crisis management efforts—such as the need for organizations to promote goodwill and responsibility for improving the community in community relations activities (Grunig & Hunt, 1984; Heath & Ni, 2010; Lukaszewski, 2013).

It is unfortunate that available texts tend to see community relations efforts as serving a different goal instead of recognizing its role in maintaining organization engagement (Heath & Ni, 2010). In essence, community relations efforts foster engagement of organizations (traditional or otherwise) with local consumers, citizens, and other organizations. Hence, we view community relations programs as crucial to CCC activities and to roles within the community. For instance, the nature of crises, risks, emergencies, and issues necessitate appropriate community relations programs. Specifically, a crisis management institution such as the CCC should not act merely as a policing agency, but should also move to engage organizations operating in high-risk industries in a way that they can mutually develop solutions or prevention plans. Such an engagement might help representatives of such organizations in their own community relations programs to be willing to disclose dangerous materials to the CCC and perhaps to seek help in the best strategies to educate the public and perhaps prepare local citizens ahead of emergencies or crises. While Heath (2005) did not specifically deal with the CCC, he alluded to the idea by stressing the virtue of community advisory committees along with other communication avenues in communities where reasonable levels of risk exist (Heath, 2005; Heath & Ni, 2010).

Furthermore, Neff (2005) presented community relations as an idea that represents a strategic opportunity to benefit multiple parties. For instance, she suggested that when community relations efforts utilize the power of communication with the goal of creating dialogue, the process is usually grounded in participants' rights and responsibilities through the use of public opinion. In a similar manner, the CCC can invoke community relations efforts that are grounded as a grassroots movements that create a unity of effort in becoming good citizens within the community. At the same time, community relations campaigns from other organizations would require commitment through active budgeting and community engagement activities that show the CCC and local residents that the organization wants to play a key role in the lives of members of the communities where it operates (opportunity to give back). This level of commitment also presents an organization with a tangible opportunity to assess and perhaps to understand the expectations that local residents have in terms of expected actions, policies, and commitment (Heath & Ni, 2010). In short, community relations efforts are considered as a valuable tool for crisis preparation and management. Heath and Ni (2010) put it this way:

> CR [Community relations] can and probably does blend with strategic skills in crisis communication, issues management, and risk communication. This arrangement is designed so that the corporate communication and management cadre can have strategic dialogue with individuals who have interests in and special needs because of the nature of local operations. (p. 560)

To this end, the different types of community relations activities have been presented (see Heath & Ni, 2010). The authors offered them as archetypal of what to expect in community relations efforts, but they cautioned that not all types of community relations activities need to be in place in an attempt to have effective results. Instead, different types of community relations efforts need to be matched with specific conditions or scenarios (Heath & Ni, 2010; see also Bitzer, 1968, 1987).

We do not want to present community relations as a panacea for crisis management. Instead, we want to stress the important role that community relations efforts can play in risks, issues, and crisis management. After all, just being a good neighbor at the expense of addressing specific underlying crisis issues may not suffice and is not a substitute. For instance, when community relations activities are used by an organization mainly as a ploy to expand its marketing or customer base, it will not work and may have counterproductive effects bound to have significant negative consequences or set crises in motion in the long term.

However, when an organization teamed up with a CCC to develop a campaign that teaches local residents the importance of understanding and dealing with certain dangerous chemicals and how to prevent children and adults alike from being injured by such chemicals, it is bound to have a positive impact on such organizations. For example, oil- and gas-fracking efforts are the rage or new buzzword in the West Texas Permian Basin and in other parts of the country dealing with shale oil and gas. However, residents within these communities are concerned about water safety issues, especially given the fact that oil companies are reluctant to disclose chemical compositions of their fracking fluids. Some of the companies are claiming that they work within the guidelines and regulations of the industry, but their statements do little to allay the concerns of residents. It might help for these high-risk organizations to collaborate with the local CCC and government agencies, such as FEMA and the CDC, on how to educate the public about their operations while providing guidelines for dealing with possible accidents. This approach to community relations efforts helps demonstrate the fact that the point of good community relations is to actively engage, participate, and contribute to a society in a positive way (Heath & Ni, 2010; Lukaszewski, 2013). Similarly, this approach to community relations efforts helps in dealing with issues of identification (Burke, 1969; Heath & Ni, 2010)—where an organization determines the underlining issue and then decides to do something about it.

Therefore, it stands to reason that any proposed community relations efforts and programs need to be designed or prepared with greater sensitivity to issues of culture and diversity. Culture is an important part of any organization (Bechler, 1995, 2004; Miller, 2014), as well as any community. Waymer (2012) reminds us that culture has a profound ramification on attitudes and beliefs of organizational members. Every organization has a story with regard to how members of the community view such organizations and its efforts. For instance, an organization that has a past record of discrimination and prejudices to certain groups is less likely to be trusted. More importantly, such organizations run a risk of the prejudicial issue becoming a crisis manifestation. Therefore, it is poignant that organizations with a tarnished or less-than-stellar record with the issue of diversity first acknowledge the problem and then begin to put in place efforts or plans that demonstrate renewed commitment to the issue of fairness and diversity. Otherwise, any effort regarding community relationships is more likely to be met with resistance and perhaps protest, which may snowball into a new crisis altogether. Consequently, vigilance through self-reflection by organizations can make the difference in overcoming crises.

Given diversity existing in most communities, it is therefore prudent that the CCC attempt to be inclusive in its approach to crisis and disaster management. Emphasis on offering consistent but multiple accounts of what happened and communicating expectations is considered an attempt to add value to society's discourse, especially in assuring different groups that their concerns will be listened to and addressed appropriately (Hon & Brunner, 2014; Waymer, 2012). Waymer (2010) also suggested that dialogue about diversity needs to be ongoing and that every attempt must be made to make sure it is meaningful. Specifically, the need to have access to risk and crisis information from sources similar to the targeted group and provided in a way that is sensitive to them will allow people to feel more prepared or at least motivate them to deal with crises and cooperate with emergency response (Waymer, 2010). Earlier, we recommend the need to include minority group leaders as crisis communication messengers in disseminating risk and crisis messages to diverse groups. Minority group leaders, including local representatives and church leaders, are ideal for serving as the messengers or liaisons between a group and crisis managers or organizations. In this regard, attention to diverse and vulnerable populations can be achieved in terms of holding special meetings and outreach programs specifically tailored to the needs of members in such groups (Veil, Mitcham, & Sellnow, 2012). In essence, we need more research that addresses inclusion, diversity sensitivity, and even need assessment in diverse populations in an attempt to not only theorize but also to have a functional application of crisis messages that work in a diverse populations.

Lukaszewski (2013) offers crisis practitioners good advice. He reminded us that an effective crisis management requires the recognition that crises create victims, thus any crisis plan must be victim- or *people-centered* (italics are mine). Furthermore, he recommended that organizations charged with crisis management go above and beyond expectation (i.e., 120% in every direction) to address victims' needs. In that public memory may be short-lived but victims' and survivor's memory long-lived, the only way to mitigate a crisis impact or to prevent it is to be sensitive to victims' needs (e.g., prompt compensation, with the proper empathy, apology, or regret communicated). At the same time, Lukaszewski (2013) provided the adaptation of the International Rotary Four-way Test for communication during crisis:

1. Is it the truth?
2. Is it fair?
3. Will it build goodwill?
4. Will it be beneficial to all concerned? (p. 334)

The CCC can benefit by incorporating these ideas into its communication and crisis plan.

Conclusion

There is much to be said about crisis planning in crisis preparedness and in the ultimate control over crises, disasters, terrorism, and emergencies. Thus, effectiveness of any crisis planning is rooted in the attempt made to plan for the unthinkable and the anticipation of the inevitable. Quarantelli (2005), among other researchers, reminded us that the planning and managing principles that hold for major disasters are also useful for catastrophes. Thus, the onus is on crisis managers and community leaders to recognize that there is no substitute for adequate crisis management plans. In essence, crisis planning and crisis preparedness must focus on keeping crisis action plans and activities as close as possible to everyday routines such that by the time a crisis occurs, individuals know how to act and perhaps help others in coping with the crisis event (Olaniran & Williams, 2001, 2012). Furthermore, it has been suggested that the less community members and residents have to worry about acting in an unfamiliar way, the better their crisis response will be (Olaniran & Williams, 2001; Quarantelli, 2005). At the same time, the CCC may make its greatest impact in crisis planning or in crisis preparation by planning from the ground up rather than from the top down. It must make every effort to mobilize grassroots efforts by fostering inclusiveness. Below is a list of websites that demonstrate how the Centers for Disease Control put together programs or information to help cruise passengers, cruise lines, and port cities (stakeholders' preparation for crisis) regarding illness outbreak (i.e., contagious norovirus).

Additional Resources

- http://www.cdc.gov/norovirus/
- http://www.cdc.gov/nceh/vsp/public/public.htm

Sample Case

The Elk River Chemical Spill Health Effects Findings of Emergency Department Record Review, April 2014

Source: Collaborative investigation by the West Virginia Bureau for Public Health (WVBPH) and the Agency for Toxic Substances Disease Registry (ATSDR), http://emergency.cdc.gov/chemical/MCHM/westvirginia2014/ in dex.asp

Background

At 6:00 p.m. on January 9, West Virginia American Water (WVAW) issued a "do not use" order. The West Virginia Poison Center started receiving phone calls from people reporting rashes, nausea, vomiting, diarrhea, and other symptoms. Emergency Departments (EDs) started seeing an increase in visits, and the WVBPH began counting the number of ED visits on January 10.

MCHM is a chemical that can form bubbles like soap to help separate coal from other rocks and minerals. This process reduces air pollution caused by burning coal. Few studies on MCHM exist and most have been conducted on animals. MCHM has been tested on rats and guinea pigs. Exposure to liquid MCHM can cause skin and eye irritation, vomiting, and diarrhea. MCHM vapors in the air can also irritate the eyes, nose, throat, and lungs. When laboratory animals are exposed at high doses, MCHM has been shown to cause problems with the liver, kidneys, blood, and the brain.

On January 21, 2014, state officials learned that another material was part of the chemical release that occurred on January 9, 2014. A smaller amount (7% by volume) of a second chemical—propylene glycol phenyl ether (PPH)—was in the same tank and entered the water system at the same time as MCHM. Health effects of PPH are similar to those caused by MCHM.

Public health officials at WVBPH wanted to understand why people were going to the ED and whether any of the illnesses reported was serious. WVBPH asked epidemiologists at the Agency for Toxic Substances and Disease Registry to help with the investigation. An epidemiologist is a public health scientist who tries to understand how and why illnesses occurs so illnesses can be stopped.

How the investigation was done

Epidemiologists from WVBPH and ATSDR put together a list of public health questions they wanted to answer. They wanted to know the kinds of symptoms people had and how they were exposed to the water. They also wanted to know what kind of treatment patients received and whether they had to be hospitalized. Epidemiologists listed all these questions on a form.

Hospitals were asked to release records for people who went to the ED between January 9 and 23, 2014, and reported illnesses related to the chemical spill. A total of 584 records were released to WVBPH for review. Using the form that included the public health questions, teams of WVBPH and ATSDR epidemiologists and nurses looked through each record to find answers to the questions.

What the investigation found

- 369 records were included in the final analysis; these records were for patients who had symptoms and reported they were exposed to the water
- 215 records were *not* included in the final analysis because:
- 41 people left the ED without being seen by a physician
- 110 ED records did not record exposure to the contaminated water; 45 persons were given a diagnosis (e.g., influenza, strep throat, scabies, and shingles) that was considered a more likely explanation for their illnesses
- 3 persons had no symptoms of illnesses recorded in the record
- 16 persons were duplicates, visiting the ED for the second or third time
- The number of ED visits went down during the second week after the chemical spill. In the graph below; the date of visit is shown on the X axis along the bottom of the graph. The number of visits is shown on the Y axis along the left side of the graph. Persons admitted to the hospital are shown in blue and persons treated and released are shown in pink.

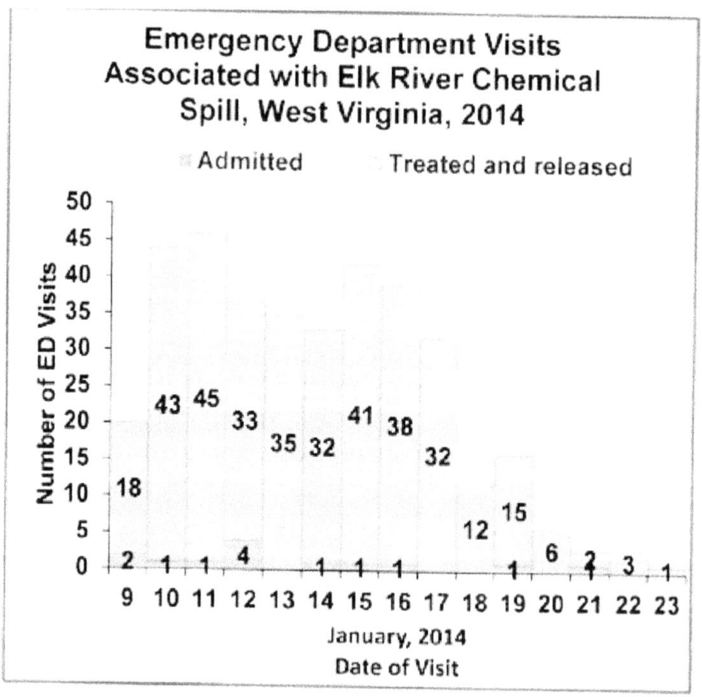

- 13 (3.5%) of 369 persons were hospitalized. People who were admitted had chronic illnesses such as kidney, liver, or lung diseases.
- 356 (96.5%) of 369 persons were treated in the ED and released. Some treatments included IV fluids and/or medications for nausea or itching.
- The most common way people were exposed to the water was bathing, showering, washing hands, or other skin contact (see Table 1).

Table 1: Exposures to Water Reported by ED Patients, Elk River Chemical Spill, Charleston, West Virginia, January 2014

Route*	Number	(%)
Bathing, showering, Other skin contact	194	(52.6)
Eating, drinking, swallowing	162	(43.9)
Breathing mist or vapor	54	(14.6)

*Patients could have more than one route of exposure recorded in the ED record.

- The most common symptoms reported were nausea, rash, vomiting, abdominal pain, and diarrhea (see Table 2).

Table 2: Most Commonly Reported Symptoms in ED Patients, Elk River Chemical Spill, Charleston, West Virginia, January 2014

Symptom*	Number	(%)
Nausea	141	(37.9)
Rash	105	(28.5)
Vomiting	104	(28.2)
Abdominal pain	90	(24.4)
Diarrhea	90	(24.4)
Headache	81	(21.9)
Itching	73	(19.8)
Sore throat	55	(14.9)
Eye pain	54	(14.6)
Cough	47	(12.7)

* Patients could have more than one symptom

- Results of laboratory tests done in the ED did not indicate any people had new kidney or liver damage.
- People who reported that they swallowed contaminated water or food, were more likely to report gastrointestinal symptoms such as nausea, vomiting, and diarrhea. People who reported skin contact with contaminated water were more likely to report redness or itching of the skin.
- What does this information mean?
- Symptoms associated with exposure to low levels of MCHM in this public water system appeared to be mild and resolved with no or minimal treatment, such as IV fluids after episodes of vomiting or diarrhea and/or medications to relieve nausea or itching.
- Most people who reported illnesses associated with the Elk River chemical spill were treated for their symptoms and released.
- Common symptoms included nausea, vomiting, diarrhea, skin rash, itching, headache, sore throat, and cough.
- These symptoms are consistent with known health effects of MCHM and with data reported by the West Virginia Poison Center. It was possible that the symptoms reported to be caused by exposure to MCHM could have been caused by other mild clinical viral infections such as colds or flu..
- There are no laboratory tests or combination of signs and symptoms that can reliably distinguish mild illnesses caused by exposure to MCHM from other mild illnesses.
- These data cannot prove that MCHM caused the reported symptoms; however, these data are consistent with what is known about MCHM from animal studies.

Next Steps

- WVBPH collaborated with the Centers for Disease Control and Prevention (CDC) on a household survey known as a Community Assessment for Public Health Emergency Response (CASPER) to measure the health, economic, and other impacts of this incident on the community as a whole. The CASPER was conducted April 8–10, 2014, and findings will be released after an analysis is complete. The number of persons seen in the ED (369) is a small proportion (0.12%) of the estimated 300,000 persons affected by the chemical spill. This household survey will be helpful in expanding the understanding of the impact on the entire population.

- WVBPH epidemiologists will use the information from the ED medical record review and the CASPER to make recommendations to strengthen emergency response in the future.
- Epidemiologists will continue to evaluate this data and finalize a scientific paper.
- Writing and talking about scientific findings will help everyone better understand all the issues surrounding this investigation.

Discussion Questions:

1. How would you assess or rate WVPC and WVBPH crisis readiness or preparedness? Why?
2. How about their response?
3. In collaboration with the CDC, what could it have been? Why or why not?
4. What recommendation or changes would you make to address similar crises?

References

An, S. K., & Cheng, I. H. (2010). Crisis communication research in public relations journals: Tracking research trends over thirty years. In *The handbook of crisis communication* (pp. 65–90). New York: Wiley.

Bechler, C. (1995). Looking beyond the immediate crisis response: Analyzing the organizational culture to understand the crisis. *Journal of the Association for Communication Administration*, *1*, 1–17.

Bechler, C. (2004). Reframing the organizational exigency: Taking a new approach in crisis research. In D. P. Millar & R. L. Heath (Eds.), *Responding to crisis: A rhetorical approach to crisis communication* (pp. 63–74). Mahwah, NJ: Lawrence Erlbaum.

Bitzer, L. F. (1968). The rhetorical situation. *Philosophy & Rhetoric*, *1*(1), 1–14.

Bitzer, L. F. (1987). Rhetorical public communication. *Critical Studies in Mass Communication*, *4*(4), 425–428.

Boyd, D. M, & Ellison, N. B. (2008). Social network sites: Definition, history, and scholarship. *Journal of Computer-Mediated Communication*, *13*, 210–30.

Brodbeck, L. (2015, July 22). Jeep hijack brings IoT cybersecurity back into focus. Retrieved from http://www.benzinga.com/news/15/07/5692329/jeep-hijack-brings-iot-cybersecurity-back-into-focus.

Burke, K. (1969). *A rhetoric of motives*. Berkeley: University of California Press.

Burnett, J. J. (1998). A strategic approach to managing crisis. *Public Relations Review, 24*(4), 475–85.

Cloudman, R., & Hallahan, K. (2006). Crisis communications preparedness among U.S. organizations: Activities and assessments by public relations practitioners. *Public Relations Review, 32*, 367–76.

Coombs, W. T. (2000). Designing post-crisis messages: Lessons for crisis response strategies. *Review of Business, 21*(3), 37–49.

Coombs, W. T. (2009). Conceptualizing crisis communication. In R. L. Heath & H. D. O'Hair (Eds.), *Handbook of crisis and risk communication* (pp. 100–119). New York: Routledge.

Coombs, W. T. (2014). *Ongoing crisis communication: Planning, managing, and responding*. Thousand Oaks, CA: Sage.

Cutler, A. (2004). Methodical failure: The use of case study method by public relations researchers. *Public Relations Review, 30*, 365–75.

Diaz, M. (2013, August 21). West Texas city still reeling from plant shutdown. From http://www.wfaa.com/news/texas-news/plainview-220600801.html

Dillard, J. P., Shen, L., & Vail, R. G. (2007). Does perceived message effectiveness cause persuasion or vice versa? 17 consistent answers. *Human Communication Research, 33*, 467–88.

Distaso, M. W., & Stacks, D. W. (2010). The use of research in public relations. *The Sage handbook of public relations* (pp. 325–38). Thousand Oaks, CA: Sage.

Dorasamy, M., Raman, M., & Kaliannan, M. (2013). Knowledge management systems in support of disasters management: A two decade review. *Technological Forecasting and Social Change*, 80(9), 1834–53.

Fink, S. (1986). *Crisis management: Planning for the inevitable*. New York: AMACOM.

French, A., Greenhalgh, C., Crabtree, A., Wright, M., Hampshire, A., & Rodden, T. (2006). Software replay tools for time-based social science data. *Proceedings of the International Conference on e-Social Science*, June 28–30, Manchester, UK.

Gilpin, D. R., & Murphy, P. J. (2008). *Crisis management in a complex world*. Oxford: Oxford University Press.

Gonzalez-Herrero, A., & Pratt, C. B. (1996). An integrated symmetrical model for crisis communication management. *Journal of Public Relations Research*, 8(2), 79–105.

Greenhalgh, C., French, A., Tennent, P., Humble, J., & Crabtree, A. (2007). From replay tool to digital replay system. *Proceedings of the International Conference on e-Social Science*, October 7–9, Ann Arbor, MI.

Grunig, J. E., & Hunt, T. (1984). *Managing public relations* (pp. 21–23). New York: Holt, Rinehart and Winston.

Heath, R. L. (2004). Telling a story: A narrative approach to communication during crisis. *Responding to crisis: A rhetorical approach to crisis communication* (pp. 167–87). Mahwah, NJ: Lawrence Erlbaum.

Heath, R. L. (2005). A crisis management perspective of business continuity, in: A. Hiles, P. Burnes (eds.), *The definitive handbook of business continuity management* (pp. 47–58). Hoboken, NJ: Wiley & Sons Ltd.

Heath, R. L. (Ed.). (2010). *The Sage handbook of public relations*. Thousand Oaks, CA: Sage.

Heath, R. L. (2012). Introduction. In *Handbook of crisis communication* (pp. 1–14). Malden, MA: Wiley-Blackwell.

Heath, R. L., & Coombs, W. T. (2006). *Today's public relations: An introduction*. Thousand Oaks, CA: Sage.

Heath, R. L., Lee, J., & Ni, L. (2009). Crisis and risk approaches to emergency management planning and communication: The role of similarity and sensitivity. *Journal of Public Relations Research, 21*(2), 123–41.

Heath, R. L., & Millar, D. P. (2004). A rhetorical approach to crisis communication: Management, communication processes, and strategic responses. *Responding to crisis: A rhetorical approach to crisis communication* (pp. 1–17). Mahwah, NJ: Lawrence Erlbaum.

Heath, R. L., & Ni, L. (2010). Community relations and corporate social responsibility. *The Sage handbook of public relations* (pp. 557–68). Thousand Oaks, CA: Sage.

Heitmann, H., & Lott, B. (2008). Protecting corporate reputation in an ERA of instant transparency. In P. Dalaleo & N. Pal (Eds.), *From strategy to execution: Turning accelerated global change into opportunity* (pp. 237–57). Berlin: Springer.

Herbane, B. (2013). Exploring crisis management in UK small and medium-sized enterprises. *Journal of Contingencies and Crisis Management, 27*(2), 82–95.

Hinduja, S., & Patchin, J. W. (2008). Personal information of adolescents on the Internet: A quantitative content analysis of Myspace. *Journal of Adolescence, 31*, 125–46.

Hon, L., & Brunner, B. (2014). Measuring public relationships among students and administrators at the University of Florida. *Journal of Communication Management*, 6, 227–38.

Jaques, T. (2007). Issue management and crisis management: An integrated, non-linear, relational construct. *Public Relations Review*, 33(2), 147–57.

Jaques, T. (2009). Issue and crisis management: Quicksand in the definitional landscape. *Public Relations Review*, 35(3), 280–86.

Jaques, T. (2012). Issue management as strategic aspect of crisis prevention. In. B. Olaniran, D. Williams, & T. Coombs (Eds.), *Pre-crisis planning, communication, and management: Preparing for the inevitable* (pp. 17–36). New York: Peter Lang.

Jin, Y., Pang, A., & Cameron, G. (2012). Pre-crisis threat assessment: A cognitive appraisal. In In. B. Olaniran, D. Williams, & T. Coombs (Eds.), *Pre-crisis planning, communication, and management: Preparing for the inevitable* (pp. 125–146). New York: Peter Lang.

Kramer, N. C., & Winter, S. (2008). Impression management 2.0: The relationship of self-esteem, extraversion, self-efficacy, and self-presentation within social networking sites. *Journal of Media Psychology*, 20(3), 106–16.

Lerbinger, O. (1997). *The crisis manager*. Mahwah, NJ: Lawrence Erlbaum.

Low, L. S., Chung, V. T., & Pang, A. (2012). Efficacy of crisis management plans: Toward a revised synthesis model in managing crises. In B. A. Olaniran, D. E. Williams, & W. T. Coombs (Eds.), *Pre-crisis planning, communication, and management: Preparing for the inevitable* (pp. 171–97). New York: Peter Lang.

Lukaszewski, J. E. (2013). *On crisis communication: What your CEO needs to know about reputation, risk and crisis management*. Brookfield, CT: Rothstein Associates.

Marra, F. J. (2004). Excellent crisis communication: Beyond crisis plans. In Millar, D. P. & Heath R. L. (Eds.), *Responding to crisis: A rhetorical approach to crisis communication* (pp. 311–25). Mahwah, NJ: Lawrence Erlbaum.

Mazer, J. P., Murphy, R. E., & Simonds, C. J. (2007). I'll see you on Facebook: The effects of computer-mediated teacher self-disclosure on student motivation, active learning, and classroom climate. *Communication Education*, 56, 1–17.

McConnell, A., & Drennan, L. (2006). Mission impossible? Planning and preparing for crisis. *Journal of Contingencies & Crisis Management*, *14*(2), 59–70.

Michaelson, D., & Stacks, D. (2010). *A practitioner's guide to public relations research, measurement and evaluation*. New York: Business Expert Press.

Miller, K. (2014). *Organizational communication: Approaches and processes*. Independence, KY: Cengage Learning.

Neff, B. D. (2005). Community relations. In R. Heath (Ed.), *Encyclopedia of public relations* (pp. 174–77). Thousand Oaks, CA: Sage.

Nichols, R. (2010, June 7). *Emergency text messaging signals evolution in public safety communication*. Retrieved from http://www.emergencymgmt.com/safety/Emergency-Text-Messaging-Public-Safety-Communication.html

Olaniran, B. A., & Scholl, J. C. (2014). New England Compounding Center meningitis outbreak: A compounding public health crisis. *Journal of Risk Analysis and Crisis Response*, *4*(1), 34–42.

Olaniran, B. A., & Williams, D. E. (2001). Anticipatory model of crisis management: A vigilant response to technological crises. In R. L. Heath (Ed.), *Handbook of public relations* (pp. 487–500). Thousand Oaks, CA: Sage.

Olaniran, B. A., Scholl, J. C., Williams, D. E., & Boyer, L. (2012). Johnson and Johnson phantom recall: A fall from grace or a re-visit of the ghost of the past. *Public Relations Review*, *38*(1), 153–55.

Olaniran, B. A., & Williams, D. E. (2012). The need for an anticipatory perspective in crisis communication. In B. A. Olaniran, D. E. Williams, & W. T. Coombs (Eds.), *Pre-crisis planning, communication, and management: Preparing for the inevitable* (pp. 1–15). New York: Peter Lang.

Oppe, S. (1988). The concept of risk: A decision theoretic approach. *Ergonomics*, *31*(4), 435–40.

Palen, L., Vieweg, S., Liu, S. B., & Hughes, A. L. (2009). Crisis in a networked world features of computer-mediated communication in the April 16, 2007, Virginia Tech Event. *Social Science Computer Review*, *27*(4), 467–80.

Palenchar, M., & Freberg, K. (2012). Emergency management planning: Risk, crises, issues, and social media. In B. Olaniran, D. Williams, & W. Coombs (Eds.), *Pre-crisis planning, communication and management: Preparing for the inevitable* (pp. 147–70). New York: Peter Lang.

Pang, A., Cropp, F., & Cameron, G. T. (2006). Corporate crisis planning: tensions, issues, and contradictions. *Journal of Communication Management*, *10* (4), 371–89.

Perry, R., & Quarantelli, E. (2005). *What is a disaster? New answers to old questions*. Philadelphia: Xlibris.

Quarantelli, E. L. (2005). Catastrophes are different from disasters: Some implications for crisis planning and managing drawn from Katrina. In *Understanding Katrina: Perspectives from the social sciences*. https://unitedsikhs.org/katrina/catastrophes_are_different_from_disasters.pdf

Scholl, J., Williams, D., & Olaniran, B (2005). Preparing for terrorism: A rationale for crisis communication center. In H. O' Hair., R. Heath & G. Ledlow (Eds.), *Community preparedness and response to terrorism* (pp. 243–68). Westport, CT: Praeger.

Seeger, M. W., Sellnow, T. L., & Ulmer, R. R. (2003). In *Communication and organizational crisis* (pp. 163–184). Westport, CT: Praeger.

Seeger, M. W., Sellnow, T. L., & Ulmer, R. R. (2012). Communication, organization, and crisis. *Communication Yearbook, 21*, 231.

Shields, K. (2015). Cybersecurity: Recognizing the risk and protecting against attacks. *North Carolina Banking Institute Journal, 19*, 345.

Shrivastava, P., Mitroff, I., & Alpaslan, C. M. (2013). Imagining an education in crisis management. *Journal of Management Education, 37*(1), 6–20.

Shrivastava, P., & Statler, M. (Eds.). (2012). *Learning from the global financial crisis creatively, reliably, and sustainably*. Palo Alto, CA: Stanford University Press.

Smith, D., & Fischbacher, M. (2009). The changing nature of risk and risk management: The challenge of borders, uncertainty and resilience. *Risk management, 11*(1), 1–12.

Stacks, D. W. (2006). Dictionary of public relations measurement and research. *Institute for Public Relations*. http://www. instituteforpr. *org*/research_single/dictionary_public_relations

Stacks, D., & Michaelson, D. (2010). *A practioner's guide to public relations research, measurement and evaluation*. New York: Business Expert Press.

Stephenson, D., & Bonabeau, E. (2007, February). Expecting the unexpected: The need for a networked terrorism and disaster response strategy. *Homeland Security Affairs*, 3(1). http://www.hsaj.org

Strong, J. T., Anderson, R. E., & Dubas, K. M. (1993). Marketing threat appeals: A conceptual framework and implications for practitioners. *Journal of Managerial Issues*, 5(4), 532–46.

Traynor, P. (2008, September). *Characterizing the limitations of third-party EAS over cellular text messaging services*. Atlanta: Georgia Institute of Technology.

Veil, S. R., Mitcham, C., & Sellnow, T. L. (2012). Lost in translation: How emergency managers communicate preparedness with vulnerable populations. In B. A. Olaniran, D. E. Williams, & W.T. Coombs (Eds.), *Pre-crisis planning, communication, and management: Preparing for the inevitable* (pp. 201–25). New York: Peter Lang.

Wallace, M., & Weber, L. (2010). *The disaster recovery handbook: A step-by-step plan to ensure business continuity and protect vital operations, facilities, and assets*. New York: AMACOM.

Walther, J. B. (2007). Selective self-presentation in computer-mediated communication: Hyperpersonal dimensions of technology, language, and cognition. *Computers in Human Behavior, 23*, 2538–57.

Walther, J. B., Van Der Heide, B., Kim, S. Y., Westerman, D., & Tong, S. T. (2008). The role of friends' appearance and behavior on evaluations of on Facebook: Are we known by the company we keep? *Human Communication Research*, *34*, 28–49.

Waters, R. D., Burnett, E., Lamm, A., & Lucas, J. (2009). Engaging stakeholders through social networking: How nonprofit organizations are using Facebook. *Public Relations Review*, *35*, 102–6.

Waymer, D. (2010). Does public relations scholarship have a place in race? In R. L. Heath (Ed.,) *The Sage handbook of public relations* (pp. 237–46). Thousand Oaks, CA: Sage.

Waymer, D. (2012). Crisis management and communication: Pre-crisis preparation that is sensitive to diverse populations. In B. A. Olaniran, D. E. Williams, & W. T. Coombs (Eds.), *Pre-crisis planning, communication, and management: Preparing for the inevitable* (pp. 281–98). New York: Peter Lang.

Welch, K. S. (2013, February 11). Cargill plant ripple effect: Second Plainview facility shuttered in wake of plant closure. http://amarillo.com/news/local-news/2013-02-11/second-plainview-facility-shuttered-wake-cargill-plant-closure

Zhao, S., Grasmuck, S., & Martin, J. (2008). Identity construction on Facebook: Digital empowerment in anchored relationships. *Computers in Human Behavior*, *24*, 1816–36.

Zhao, W., & White, G. (2014, January). Designing a formal model facilitating collaborative information sharing for community cyber security. In *System Sciences (HICSS), 2014 47th Hawaii International Conference on System Sciences* (pp. 1987–1996). Institute of Electrical and Electronic Engineers.

PART II: PRACTICAL ASPECTS

· 6 ·

HEALTH CRISES: EPIDEMICS AND BIOLOGICAL IMPACTS

Chapter Highlights

- Types of biological and health-related threats
- The impacts of biological and health-related threats
- Improved infrastructure, outreach efforts, and other strategies of the CCCs
- The CCC's role in information dissemination, coordination of local organizations, and relating to the media

Threats to a community's biological health and safety—real or anticipated—deserve serious attention from the CCC. The high price of unpreparedness for biological threats can include litigation and court costs, damage to property and health, and the loss of the public's trust (Scholl, Williams, & Olaniran, 2005). Community members and leaders can never fully anticipate such costs, which are likely greater than those incurred when preventive measures are taken.

The issue of health and biological crises deserves special attention in this chapter. As explained later on, crises that affect public health don't necessarily have to originate from health threats, but they can also emerge from nonhealth disasters, such as terrorist attacks and naturally occurring events like earthquakes. Public health and safety encompasses many aspects: biology, environment, lifestyle, and health services. Virtually every organization and service provider within a community has a stake in public health. The success of local actions to address and to curtail biological disasters rests in the ability of community and municipal agencies to coordinate and to be in communication with one another. A discussion and application of the anticipatory model for crisis management are appropriate when it comes to health and safety threats. For example, local health communities have to be in a vigilant and anticipatory mode in order to expand and to maintain their treatment capabilities. They also need to take steps to facilitate stress management and to provide psychological support, and put in place protocols to contain harmful biologic threats.

One common shift in the public health sector has been improved communication between medical personnel and first responders. A CCC at the local level can play a pivotal role in helping a community become better prepared to handle health-related disasters. The CCC can make sure the appropriate agencies communicate with each other as well as provide venues at the CCC and elsewhere to strengthen these connections.

This chapter overviews the common types of biological and health-related crises, such as bioterrorism, epidemics, and impacts of organizational misconduct. Next, the impacts and outcomes of various biological crises are discussed. The chapter outlines strategies that can help address the most common health disasters, such as improved infrastructure and vaccination campaigns, and how CCCs can help design and implement these strategies. The CCC's role in information collection and dissemination as well as its coordination with local media is also examined.

Health-Related Crises and Their Impacts

Biological diseases: There is a distinction between biological disasters (e.g., flu epidemics) and natural disasters (e.g., hurricanes), even though the two are usually linked. While this chapter does not directly address the environmental impacts of hurricanes, earthquakes, or other natural events, attention should be paid to their health-related implications. For instance, the 7.0 earthquake that hit Haiti in January 2010 indirectly triggered a nationwide cholera epidemic. The country saw new cholera cases as recently as March 2012, which marked the start of the rainy season (Associated Press, 2012).

Also noteworthy are the thousands of individuals given antibiotics in the wake of the 2001 anthrax scare in the United States (Barbisch, 2005), and the threat of nuclear fallout from the biological weapons being developed by many nations. Additionally, community leaders' and experts' responses to biological disasters are similar to those executed during natural catastrophes. Lerbinger (2012) states that both kinds of disasters require that responders lessen a community's vulnerability and already have in place measures and warning systems that reduce the harm: "However, the duration of a biological crisis lasts for months and years while that of a natural disaster is measured in days and weeks" (p. 114).

Epidemics and pandemics: The most feared type of biological threat is a pandemic (Lerbinger, 2012), an infectious disease that spreads rapidly across regions and countries throughout the world. According to Lerbinger (2012), a common characteristic of pandemic-level disease is the transfer of infection from animals to humans, which often comes from shifts in the population from rural to crowded urban areas. This was the case for the outbreak of influenza and other infectious diseases in the United States in the early twentieth century (Zinn, 2003). Another factor is the increase in industrial farming. The mass production and crowding of chickens, pigs, and other animals can help cultivate bacteria that are difficult to fight and to control (Lerbinger, 2012).

A more recent biological threat has been the 2014 outbreak of the Ebola virus in the western African countries of Guinea, Sierra Leone, and Liberia (CDC, 2015). Four cases were confirmed in the United States in September and October 2014, one of which resulted in death (CDC, 2014). Despite the relatively low numbers of cases in the United States, national and local medical centers and authorities were constantly under the public eye regarding their response to, and mitigation of, these cases. When cases arose in Texas Presbyterian Hospital in Dallas, many concerned citizens scrutinized the hospital and local officials for their handling of the cases. Texas Presbyterian was also criticized for its alleged inadequate training of medical staff in contact with infected patients, especially those who cared for Thomas Eric Duncan, a Liberian citizen who settled in the United States (Fox, 2014). The hospital also came under fire because one of its nurses decided to fly on a commercial airline after she has been caring for Duncan; she was later diagnosed with the virus, but survived.

As we mentioned earlier, natural disasters, such as earthquakes, usually give rise to massive amounts of injuries and widespread viral outbreaks. The Fukushima Daiichi nuclear disaster, caused by the 2011 tsunami off the

Japanese coast, created a number of issues. The failure of emergency power supplies to kick in (Nöggerath, Geller, & Gusikov, 2011) can lead to people going without basic food and necessities, which can lead to sicknesses, viral infections, and untreated injuries. The structural damage caused by the 2010 earthquake in Haiti gave rise to physical injuries and deaths. The earthquake also resulted in over 700 deaths from a cholera outbreak, which affected more than 100,000 people (Danticat, 2012).

Lerbinger (2012) reports that biological diseases are becoming major killers, and they constitute natural disasters that can threaten the physical environment. The emergence of other diseases is not just a national or international concern; such disasters have specific impacts on local communities. For instance, communities vary in their level of preparedness—in particular, the level of preparation and anticipation in which they engage—as well as the resources they have to combat biological threats. Also, some communities are more capable than others at coordinating the communication efforts among health-based agencies and with the local media. In all, these variations indicate a community's level of readiness to react quickly when a health threat or disaster strikes.

Bioterrorism: Inungu (2005) defines bioterrorism as the real or threatened use "of biological agents (bacterial, viral, parasitic, or biological toxin) to promote or spread fear or intimidation upon an individual, a specific group, or the population as a whole for religious, political, ideological, financial, or personal purposes" (p. 1). The use of biological terrorism or warfare has been documented since the sixth century BCE (Du, 2004). According to Inungu, biological agents used throughout history have included anthrax (caused by spore-forming bacteria), botulism (caused by spore-forming toxins that cause muscle paralysis), plague (found in rodents, cats, rabbits, and fleas), smallpox, ricin (derived from castor beans used to make castor oil), Salmonella poisoning, tularemia (i.e., rabbit fever), and brucellosis (caused by the *Brucella* bacterium).

Of all the tools used to inflict terror on a community, biological agents are the most accessible and potentially the most deadly. Von Lubitz and Lary (2005) point out that biological agents are easy to make, conceal, carry, and spread throughout entire populations. Unlike detonations or mass shootings, biological attacks are covert: "An act by a lone terrorist could result in a major disease outbreak that is spread in successive waves throughout the population before anyone is even aware that it has happened" (Action & Ledlow, 2005, pp. 108–9). Biological agents can incubate for several weeks, which allow bioterrorists to escape detection. In addition, first-stage symptoms of most biological agents look very similar to common illnesses, such as lung infections and influenza, which makes an attack almost impossible to identify during its early stages (Action & Ledlow, 2005, p. 109), which compromises crisis planners' ability to engage in signal detection and practice vigilance.

The United States has had relatively little experience with bioterrorism. Nevertheless, there is a rise in the number of chemical and biological weapons being produced throughout the world, and the United States is still not adequately prepared to handle such a catastrophe (Cole, 2012), partially because financial constraints make it a low priority in some communities (Action & Ledlow, 2005). Local communities should assess their levels of health and safety within the realm of terrorism. Relatively recent events like the attacks on 9/11 have placed more attention on the role of public health, particularly at local levels: "Although terrorism is an international and globalization concern, and it portend[s] broad environmental consequences, medical emergencies will [have the greatest impact] on local communities" (O'Boyle & Simms, 2005, p. 249). No matter how modest, anticipated casualties require that local health communities expand their triage and treatment capacity, take steps to provide stress management and mental health service, and put in place protocols for isolation and quarantine (Barbisch, 2005; Cole, 2012). To this end, a sustainable, comprehensive healthcare management system at the local level might improve efforts at meeting the overwhelming clinical needs that come from a widespread health disaster.

Organizational Misconduct

Biological threats also come from malpractice and carelessness of organizations that produce medical and health-related products (e.g., pharmaceuticals and over-the-counter drugs). Biological or health-related crises that stem from organizational misconduct have the potential to hinder the health and safety of employees, customers, and the community. They also threaten public trust and damage company reputations.

All too often there is a lack of attention and vigilance paid to conditions that threaten a company's ability to serve the public's safety and health needs. When it comes to crisis preparation, some company leaders and stakeholders tend to respond to crises after they occur instead of committing resources, time, and energy to preventing such catastrophic events. Unfortunately, this carefree approach is becoming more commonplace.

Johnson & Johnson and the "phantom recall": As a company that has enjoyed a long-standing history and strong image, Johnson & Johnson has been regarded in the past as the epitome of crisis management. In 1982, the company was forced to address two tampering incidents with its best-selling, over-the-counter Tylenol capsules that resulted in several deaths. In response to the 1982 incidents, Johnson & Johnson announced its withdrawal of the faulty Tylenol products from retail shelves and introduced new tamper-resistant packaging. Then, in 1986, it was reported that a New York woman died from another Tylenol tampering. The 1986 episode led to Johnson & Johnson's final removal of capsules, replacing them with caplets.

Over the years, families have come to trust Johnson & Johnson, and as a result the company has received the benefit of the doubt from the media and from the public that it will not knowingly inflict harm, even in the wake of the tampering scandals (Olaniran, Scholl, Williams, & Boyer, 2011). Burke (1999) and George and Evuleocha (2003) point out that a well-reputed company like Johnson & Johnson will often command stakeholders' and communities' willingness to forgive them when crises emerge.

Olaniran and Williams (2001), in their development of the anticipatory model, conducted an analysis of this series of tampering incidents as well as Johnson & Johnson's crisis responses. They questioned why it took Johnson & Johnson three separate incidents spanning seven years to finally remove the defective with capsules from the market. Olaniran and Williams further argued that Johnson & Johnson's decision to recall was more out of concern for the company's profits than out of a show of genuine distress for consumers. While this claim is debatable, James Burke, CEO of Johnson & Johnson at the time, later admitted that he regretted not replacing the capsules earlier than he did (Dezenhall & Weber, 2007).

The dust did not settle for long. Johnson & Johnson was again in hot water over the recall of its Children's Motrin. In April 2010, the company recalled millions of bottles of potentially contaminated over-the-counter medicines, such as Children's Tylenol and Benadryl, forcing one of its plants to shut down well into the following year and prompting civil lawsuits and a criminal probe. Contrary to the widely publicized recalls of the past, Johnson & Johnson's 2010 campaign included a quiet buyback of its Motrin painkiller by its own contractors in an attempt to avoid publicity (Gibson, 2010; Olaniran et al., 2011). It turns out that the faulty tablets were originally distributed to gasoline service stations. After the company learned the medicine wasn't dissolving quickly enough when taken orally, Johnson & Johnson hired contractors to remove the remaining tablets from the stores rather than ordering a public recall.

CCCs should play a role in monitoring the activities and practices of pharmaceutical and medical-supply companies operating in their communities. Because of their physical location, local authorities should routinely scrutinize contamination from manufacturing facilities, and the CCC can play a part in raising the red flags that signal the need to scrutinize a facility, unit, or practice. Communication experts within the CCC can help decide who in the community has the credibility to raise the flag, the content of the warning message(s), and to whom and how the message should be delivered. These considerations warrant the attention of health providers and public health professionals, who can further adapt the message to enable crisis responders to address health-related concerns within the local community. Furthermore, while the CCC is there to protect the general well-being of the community, there is nothing to say that corporations could not work alongside CCCs on potential problems if needed.

The CCC's role can also include coordinated programs (and the managers of those programs) that maintain routine quality control so that it does not take a regulatory authority to find a problem. For instance, it would appear prudent for Johnson & Johnson to have a plan of periodic evaluation of its factory for possible contaminants and optimal steps to deal with emerging problems (Olaniran et al., 2011). Company personnel might have the primary role in developing the evaluation plan, but other experts connected through the CCC can assist with packaging the plan so that a variety of professionals can easily understand and implement it. In general, these preventive measures imply that decision makers and other stakeholders should not have to wait until they are notified by an external agency that contaminants were found in its processing plant. In the end, a company of Johnson & Johnson's reputation and experience should never have allowed itself to be caught up in the phantom recall predicament (Olaniran et al., 2011).

Health-based companies like Johnson & Johnson have some responsibility of educating the public on health matters, controlling the outbreak of illness and infection, maintaining and monitoring industrial health standards, and informing the public of threats to its health and safety (Schneider, 2006). These companies also have a responsibility to be transparent about product flaws. Upholding public health implies that agencies and companies not only be reactive when a health threat is eminent, but also be proactive in protecting the public from unhealthy conditions. When companies are not able or willing to maintain these standards, they should consult other professionals who can help them accomplish their goals. The CCC is one such agency that can assist, particularly in helping the company communicate its intentions to the public and finding the best ways to respond to health-based crises as quickly as possible.

Outcomes and Impacts of Biological Threats and Disasters

Having examined the different kinds of biological threats, we also need to understand the nature and extent of the short- and long-term damage caused by these threats. The most obvious might be the spread of viral disease with devastating symptoms, the deaths of loved ones, heightened security, and the demolition of buildings and structures. Many of these effects, which might appear to be mainly political and environmental, also carry medical implications. Dead bodies that go unrecovered for an extended period of time will breed deadly bacteria and other microbial agents. Increased security by the local police and armed forces—not to mention the possibility of terrorist acts—could lead to increased distress, policy brutality, and widespread panic among the population. Also, the sudden destruction of buildings and weakening structures put people at increased risk for injuries and deaths.

It is easy to focus attention on the most immediate and visible effects, which can be devastating and large in scale. However, there are more hidden and subtle effects that cannot be ignored, such as effects on behavior and psychological health. Localized crises have a way of disrupting "communication practices at the individual, relational, and societal levels, which are closely related to … health" (Becker & Thompson, 2005, p. 56). After terrorist attacks, individuals might come to the hospital presenting rogue physical symptoms, which are often aggravated by psychological stress (Taintor, 2003). A CCC working in tandem with a federal organization, such as the Federal Crisis Counseling program (Rosen, Greene, Young, & Norris, 2010), can be better informed about the kinds of psychological stressors most likely to occur after various types of crises and help coordinate efforts to get the help individuals need.

Community-wide crises bring about many unnoticed effects on psychosocial health, especially on those who live or work relatively close to the disaster's epicenter (e.g., people who lived and worked in Manhattan during 9/11).

Many individuals who experience trauma can feel disconnected from others and have difficulty making sense of reality (Niederhoffer & Pennebaker, 2002). The purpose of terrorism is to inflict religious, financial, political, and/or environmental harm, all of which can threaten the mental health of the community affected (Becker & Thompson, 2005). In addition, people who are directly affected—as opposed to those who learn about the events after the fact—can become emotionally distant from others. Unfortunately, it is only after the trauma that people wait to seek help, even after their jobs, health, and relationships fall apart (Sheely, 2003).

The CCC can play a large part in developing one or more action plans to prepare for the psychosocial impacts of a health disaster. This means "that problems can be alleviated before they mushroom into more perilous predicaments.... Through communication campaigns....individuals can overcome obstacles to regain psychosocial health" (Becker & Thompson, 2005, p. 55). In preparation, the CCC can generate and circulate a list of medical leaders and decision makers of the local or county health departments, hospitals, clinics, and mental health agencies. CCC personnel, after having done much research on the community's vulnerabilities to health and safety, can help create common templates for public health messages and campaigns, plans of action, and emergency response protocols. These templates can be maintained by the CCC, updated on a regular basis, and made available to all public health sectors.

Caution must be maintained when utilizing media outlets to prepare for and help in the event of widespread psychological damage. Publicizing images or detailed descriptions of death and destruction in the wake of disasters might undermine efforts to preserve the public's mental health. For example, frequent observations of unsettling images have been associated with depression and post-traumatic stress disorder (Ahern, Galea, Resnick, Kilpatrick, Bucuvalas, Gold et al., 2002; Saylor, Cowart, Lipovsky, Jackson, & Finch, 2003). To prevent or to anticipate this kind of emotional impact, professionals should be ready to provide social support to individuals who view such images. Such intervention can help community members cope; even encouraging some levels of self-disclosure about traumatic events is pivotal to healing (Kulic, 2003; Pennebaker, 1997). Personnel at the CCC can keep a close watch on the media and how they are covering the events. They can also minimize the potential negative impacts by disseminating messages that provide access to social support networks and groups.

The CCC: Preparation Specific to Local Communities

There is much debate about the level of preparedness in the United States against biological attacks and disaster, in fact, most communities still fall short (Cole, 2012). The first step in moving toward enhanced preparedness is to take stock in the information that is available about a community's health risks and hazards, as well as the services and resources that are already in place. Put simply, this all boils down to the acquisition, control, and dissemination of information, which is an important part of maintaining a community's health. According to the Centers for Disease Control and Prevention (CDC, 2010) a significant part of public health is providing information that allows relevant individuals to "make the best possible decisions about [a community's] well-being, under nearly impossible time constraints, while accepting the imperfect nature of their choices" (paragraph 1).

Adequate control and use of health-related information and community-specific preparation might enable elected officials, medical decision makers, and message designers to reach their intended targets about biological disasters and risks with the most success. Knowing the most likely biological threats to a community, as well as what types of messages and message sources would be best accepted, might ensure that community members remain safe and recover from catastrophic events (Scholl, Olaniran, Williams, & Heuman, 2007).

To assist in decision making that is in the community's best interests, a process of crisis management must be in place, which includes assessment, policy development, and assurance. According to O'Boyle and Simms (2005), whereas "assessment aims at detecting and analyzing emerging trends in infectious and noninfectious disease and environmental hazards, policy development focuses on primary prevention for target populations, and assurance focuses on community capacity to prevent disease and injury" (p. 251).

The anticipatory model for crisis management can help illuminate the significance of these steps. Assessment is a pre-crisis stage, in which enactment and expectation are best utilized. Specifically, enactment and expectation refer to the actions and assumptions made about potential biological events. These actions can help assess which mechanisms or factors within a specific community might trigger a disaster. With regard to policy development, control (another component of the anticipatory model) enables decision makers to develop policies that ensure more appropriate and effective action during a crisis response. Finally, assurance provides quality control of the current policies and allows for adaptation and correction as needed. The assurance step should be fluid enough (i.e., not too rigid) to allow for the detection of signals that might fall outside the realm of standard detection strategies (e.g., a noticeable upsurge of flu-like symptoms that still fall within "acceptance" ranges). On the contrary, when the assurance system is too rigid, health professionals might ignore unique characteristics that could foreshadow a potential crisis. For example, if an unusually large number of people in a community are experiencing flu-like symptoms of fever and vomiting, crisis managers might acknowledge that standards are still being met, but they might still do some investigation, which could uncover a batch of contaminated produce that has been shipped to a local chain of grocery stores.

Improved infrastructure: Despite improvements in our public health system, there are still debates over who should carry the burden for taking care of the sick and dying, especially when needed funds, medical personnel, and other resources are still in short supply. In the case of a hepatitis A scare in Lubbock, Texas, that originated at a local restaurant, the community members who requested the vaccine were asked to pay $15 to help defray costs, since the vaccine was not mandatory (Hepatitis Attorney, 2012). Asking victims to pay for their own treatment might be telling of the diminished quality of the current health infrastructure in place. To address adequately the needs of the community, a strong infrastructure entails genuine relationships among different health sectors to attend to the varied health needs of a community. These relationships do not merely exist; they should be based on mutual respect and consideration of the community's strengths and weaknesses.

An adequate health infrastructure is crucial to anticipating and overcoming a biological crisis. The lack of hospital facilities, personnel shortages, and inadequate medicine will almost certainly exacerbate the impact of a crisis. For instance, "many victims of HIV/AIDS do not know that they have the disease because medical facilities to test for it aren't available, and even when detected and drugs are made available, their regular scheduled use is not administered or monitored by health workers" (Lerbinger, 2012, p. 124). Therefore, a lack of adequate testing in local or county health clinics or limited public health messages that encourage testing might contribute to an upsurge of infections within a community. This problem exists in Lubbock, Texas, which is known for its high rates of sexually transmitted infections (Nightingale, 2010). According to Christine Susalski (personal communication, February 29, 2012), a public health technician with the Department of State Health Services in Lubbock, the city and county lack enough facilities that offer free or inexpensive testing for sexually transmitted infections, which can contribute to the continued spread of infections across the community.

Experts affiliated with a local CCC can help improve the local infrastructure. Individuals who are trained in viral diseases can advise on how much medicine to keep on hand within the community. More specifically, the CCC can put into place a protocol that includes how much of each type of vaccine or drug to keep on hand, a protocol that emergency response and medical personnel can consult quickly. Such a procedure would be updated periodically to accommodate for fluctuations in the local population or severity of certain disease threats. We might consider the community-based initiative about which Barbisch (2005) writes—the Seamless Emergency Medical Logistics Expansion System (SEMLES) in northern Virginia and in Maryland. The SEMLES is a comprehensive program that helps all elements of the medical and health community collaborate with each other:

> SEMLES starts with a review of assets that fall within the regulatory authority of the state or territory … [and] establishes a hub that defines and coordinates existing assets, identifies gaps, defines needs, supports planning and training, and becomes operational during an event. (Barbisch, 2005, pp. 78–79)

A local CCC can serve as the main hub through which similar health collaborations can take place. The SEMLES serves as a model for how these partnerships can improve local and regional health provision and readiness, which in turn improves the overall response to unexpected crises.

Surge capacity: A better medical infrastructure can ensure that a community maximizes its surge capacity. According to the American College of Emergency Physicians (2012), surge capacity is the measure of a facility's ability to handle a sudden incursion of patients. Surge capacity is "dependent on a well-functioning incident management system and the variables of space, supplies, staff and any special considerations (contaminated or contagious patients)" (Barbisch, 2005, p. 80). In their efforts to meet quick surges in demands, hospitals, clinics, and healthcare organizations must be seamless in the way in which they integrate their services in order to provide the best collaborative responses.

Surge capacity might also apply to the availability of services that are not necessarily hospital-related, such as counseling. Post-disaster harm to the public is not always physical; victims and those affected often experience distress and psychological harm also. Certain levels of social support need to be available for the community so that they not only are able to deal with the trauma on an individual basis but can also help each other cope with difficult situations and curtail distress.

Social support ensures that individuals can remain calm and think clearly in order to remain active and contributing members of the community; in turn, these individuals can receive information accurately and help others in need. Without adequate avenues for social support, some community members might revert to rumination, disconnection, and prolonged depression. Becker and Thompson (2005) claim:

> Victims who segregate themselves and repeatedly mull over the causes, meanings, and consequences of their depressive symptoms are at risk for deteriorating psychosocial health. In contrast, a problem-solving, active coping style is directly associated with greater satisfaction with health care and less substance abuse. (pp. 58–59)

The CCC might be a physical setting in which local healthcare professionals can set up ad hoc support groups and counseling sessions for individuals who otherwise might not have access. Such services could be temporary and yet meet the surge in the demand for support and counseling. After the initial impact of the disaster has abated, the CCC can continue to link professionals together to provide continued support and to help mental health agencies publicize their services. With this in mind, the CCC will achieve the best outcomes if its experts take inventory of the services already available, recruit the best talent, and link those professionals and services to individuals in need.

Vaccination campaigns: Many local health departments conducted mass vaccination exercises in response to the 9/11 terrorist attacks. Curran (2005) writes about one such case taking place in North Carolina. The campaign's goal was to reduce the risk of an influenza outbreak, especially in high-risk populations. According to Curran, "the mass vaccination exercise brought together community health agencies and county government with the common goal of demonstrating preparedness in the event of emergency" (p. 171). Moreover, the campaign had positive impacts: increased vigilance of possible epidemics, coordinated communication and meetings among agencies, stronger relationships among the health centers and hospitals, and increased interest from the community.

Other communities can replicate such successful efforts. The development and stocking of vaccines and antimicrobial treatments is an important part of crisis response (Action & Ledlow, 2005). Even when a crisis is not taking place, organizations in the community can encourage vaccinations in order to increase their protection against infection (de Perio, Wiegand,

Brueck, & Scholl, 2014). A local CCC can be at the helm with outreach and promotional campaigns. The CCC also can coordinate the stockpile of needed treatments by designating and/or documenting representatives for each agency and by providing a neutral arena in which all representatives can meet. Local media outlets can also provide credentialing for the involved agencies and promote the campaign.

Gathering and Disseminating Locally Based Information

The treatment of information is crucial to the success of preventive measures. Kreps, Alibek, Baley, Neuhauser, Rowan, and Sparks (2005) point out that "effective communication for biodefense involves multiple challenges and competencies in information gathering and dissemination" (p. 191). In addition, such information needs to be handled by knowledgeable and credible spokespeople, who can respond to local crises and address the media and community appropriately (Scholl et al., 2007). In terms of biological threats to the community's health and safety, these crises would require experts with the proficiency to handle specific issues, for instance, contamination of the local water supply or viral outbreaks among children in the public schools.

Educating the medical community: The key to biological threat prevention is gathering much of the information that would be used to make crisis decisions and anticipating who would play what key roles during the crisis (Scholl et al., 2007). Once gathered, this information should be used to keep the medical community informed, especially because threat vulnerabilities, biological agents, and medical advancements are always changing. "The emergence of new medical threats … [introduce] a new challenge to educate the large numbers of pre-hospital and emergency room personnel needed to ensure a maximum level of readiness" (von Lubitz & Lary, 2005, p. 269).

Educating and updating the medical community involves detecting the signals of possible threats and drawing attention to them. Potential signs of a biological threat include, but are not limited to, a noticeable increase in people seeking health care, rapid increases in fatalities, an upsurge in diseases among people who go outdoors, unexplained smells, low-flying clouds or fog unrelated to the weather, and unusual numbers of dead or dying animals in the area (National Terror Alert, 2012). A CCC can examine these signs, take inventory of them, and assign the community agencies equipped to diagnose and to monitor these signs. Additionally, the CCC can help formulate a response protocol for each signal or sign for the corresponding agency. A CCC also could be connected with the CDC, the World Health Organization (WHO), and other agencies for the information they provide. The WHO's four influenza alert levels help national and regional health authorities determine how to respond to health warnings (Schnirring, 2013). This system is not without its critics: "Some authorities believe that WHO's pandemic scale sends the wrong signal to the public" (Lerbinger, 2012, p. 129), in that since the levels signal the severity of the disease, it can be misleading because not all countries or regions experience the same level of severity.

Properly educating the medical community means cooperating with medical and graduate educational institutions, particularly when it comes to correcting communication-based deficiencies:

> problems ranging from communication and definition of professional identities and roles within the medical management team … to errors in medical command of EMS operations or even fundamental inadequacies in training of pre- and post-hospital healthcare providers.… This is particularly pronounced in rural and remote regions. (von Lubitz & Lary, 2005, p. 269)

In more general terms, health professionals and educators do not necessarily have the communication skills pivotal to coordinating care among various groups. Such skills have to be taught and put into practice.

A CCC would not be directly responsible for the medical training or for facilitating the training programs that would prepare health professionals in a variety of fields. However, the CCC is in a good position to identify problems with the communication within that medical training that can hinder its effectiveness. For instance, CCC experts with knowledge in healthcare communication (particularly interpersonal and organizational areas) can inform simulation-based medical training. Von Lubitz and Lary (2005) point out:

> The use of [high-fidelity patient simulators] in hands-on training eliminates all risk factors of similar training on living patients, permits improvement of diagnostic skills by allowing practical understanding of the involved steps, hones interactions of medical teams, and promotes punishment-free learning based on one's own errors. (p. 272)

Communication experts can assist trainers in the educational and pedagogical aspects of the training, which include giving constructive feedback that is clear, easy to replicate, and does not feel punitive. The CCC might also have personnel experienced in research skills that can help trainers collect data in order to assess the effectiveness of the medical and communication-based training.

Coordination among local entities: Medical personnel and relevant decision makers have to consider whether a biological disaster will cause a disruption in utility services or halt delivery of supplies; such delays could last as long as 18 months (Lerbinger, 2012). In addition, individuals in the community might avoid certain public places because of fear of contamination; this kind of fear could also impact travel and slow down business for retailers. A contingency plan could anticipate these effects and even reduce them. Such contingency planning might be made easier if a CCC facilitated coordination among local businesses. Key decision makers from each business could gather together to receive instructions and direction from CCC personnel—in a meeting format, social media, or other synchronous channels—as well as have the opportunity to engage each other, receive cross-training, share insights, and ensure necessary redundancies in response procedures.

Such coordination still involves the artful and skillful use of information, which can take place within the CCC. Specifically, augmenting communication between medical and public health communities helps fight biological threats. Augmentation might include teaching surveillance strategies and providing reports of potential outbreaks. Strategies should also include seeking support from the CDC and other relevant agencies in formulating and implementing a localized plan. Moreover, clinical laboratories in both the private and public sectors are crucial to identifying biological agents and to preventing their spread. A local CCC can investigate efforts for establishing national laboratories and act as a conduit for connecting local health agencies with these national labs.

When agencies are coordinated, they can receive synchronized training from the CCC in signal detection (Scholl et al., 2007), empowering service providers, conflict resolution, and consensus-building (O'Boyle & Simms, 2005). While these key players are engaged with each other, they can take advantage of increasingly common educational programs. In addition, CCC personnel can facilitate a discussion on health-related ethical issues, such as local health workers' obligations when putting themselves at risk, how to allocate limited resources, restrictions on freedom and mobility, and community members' access to information (Lerbinger, 2012).

The CCC can play a pivotal role in credentialing the leaders in the public health sector so that they can appeal to a diverse local base. O'Boyle and Simms (2005) make a projection:

> In order to develop new solutions to both population-based and community health problems,... the public health leaders of tomorrow will be required to unite and work with multicultural groups. These leaders need to embody a visionary leadership style that inspires and motivates others and, in the process, seeks new solutions, culminating in a sense of shared ownership. (pp. 263–64)

The primary personnel of the CCC would ideally have academic and professional experience in organizational communication and, therefore, would be poised to coordinate such an effort.

Educating the public: The aforementioned medical training and community coordination serves what is arguably the most important end—educating and preparing the public. Enlightening the local community on disease transmission, detection, and treatment is an indispensable component of a crisis management plan. In addition, community members generally want to be notified about potential and actual threats to their health and safety so that they know what to do individually and collectively to prevent or minimize their effect (Scholl et al., 2007). At the same time, most individuals do not want to be bothered by any information that does not foster these goals (e.g., background information about the crisis responders or medical experts).

Efforts to educate the public should be intended for the community's unique characteristics. To design the most comprehensive and appropriate localized messages, CCC team members must conduct community assessments to determine the public's biggest vulnerabilities as well as what kinds of programs or campaigns work best for the targeted audiences (Scholl et al., 2007). Team members' local knowledge might help them identify the biggest biological threats, such as what kinds of bioterrorist attacks are most likely to hit the community and what areas (e.g., water supply, industries, and health facilities) would be most devastated. Following the example of Scholl et al. (2007), CCCs should administer surveys and focus group interviews to learn the best ways to design messages and prevention plans. Also, despite preferences for different media (e.g., social media, radio, or television), the CCC should use combined or multiple-media approaches to ensure that the majority of the citizens are reached.

An important source of information gleaned from the community is preference for information, specifically what types of information should go into the mediated messages. Experts also should consider to what components and media outlets community members are most likely to pay attention (Craig, Olaniran, Scholl, & Williams, 2006). For example, Scholl et al. (2007) conducted a localized study of a community's message preferences regarding terrorist threats and acts. They found that message components such as background information about the alleged terrorist(s) were deemed less significant than specific descriptions of the threat or act, information about possible harms or outcomes, and what community members should do to protect themselves.

The CCC's Relationship with the Local Media

The various media outlets available allow community members to obtain health information relatively quickly as well as to engage in public discourse about certain events impacting several aspects of their health. When it comes to alerting the public, many officials lack the efficacy to work with the media to be effective. To that end Scholl et al. (2007) suggest that spokespeople and message designers make the following considerations:

- Which media source(s) should deliver the messages?
- How will information about potential threats differ from information about actual incidents once they occur?
- When should members of the community be alerted about health and safety threats?
- Which media representatives would be best to deliver the messages?
- What specific informational components should be included in the messages?

Of all the available media, the Internet has likely had the most noteworthy influence on the way in which decision makers and public relations experts disseminate important health information, thereby creating a sense of community (Leaffer & Mickelberg, 2006). The Internet is another valuable source for health news, on which many individuals rely. The CCC personnel who are knowledgeable about computer-mediated communication and health and media literacy can provide guidance to the public so that individuals know how to be discerning consumers of this kind of information. When such personnel have to communicate with the public via the preferred local media outlets, they can supplement their messages by referring listeners to the most relevant websites. Also available are such tools as crowdsourcing and web applications (Kamel Boulos, Resch, Crowley, Breslin, Sohn, Burtner, Pike, Jezierski, & Slayer Chuang (2011) to provide easy access to planning tools and information. The media play a crucial role in mobilizing a community and educating people on how to respond to health crises. The best mobilization efforts attempt to reach out to disenfranchised and ethnic groups that might otherwise have limited access to the media: "While the trends toward increased consumer use of online health information are truly encouraging, they also reveal several key demographic differences in consumer usage of the Internet" (Leaffer & Mickelberg, 2006, p. 54). In turn, these discrepancies can affect who obtains the greatest benefit from relevant health information.

Communication experts within a CCC can help other professionals learn who is using the Internet and who has access to it. The UCLA Digital Future Report (cited in Leaffer & Mickelberg, 2006) indicates that those who have greater access to mediated health information (e.g., via the Internet) report good or excellent health. The CCC can play a crucial role in narrowing the digital divide that still exists within most U.S. communities (Brodie, Flournoy, Altman, Blendon, Benson, & Rosenbaum, 2000) and throughout the world (Dzenowagis, 2011). Experts in health campaigns and public relations should find ways to reach more marginalized demographic groups, who have limited access to medical information and thereby report lower levels of health. Such interventions might include training targeted populations in the use of publicly accessed information (Ruffin, Cogdill, Kutty, & Hudson-Ochillo, 2005). The CCC might be a venue for targeted groups to gather and to learn how to access the Internet and, more specifically, to learn how to search for, and scrutinize, medical information online in order to increase their health literacy. As another solution, New America Media (NAM) has produced an emergency network system that will utilize media outlets that cater to specific ethnic and cultural groups. Lerbinger (2012) writes about NAM, claiming, "Ethnic media reps will activate their media outlets to deliver the messages to their millions of readers, viewers, and listeners not usually reached by mainstream media" (pp. 129–30).

To summarize, public health and public relations experts must exploit the most effective strategies for improving health behaviors and beliefs among targeted communities, especially in the wake of major biological disasters. This is especially the case given the fact that "the media … propel [unintended messages and effects] along in the name of free speech, open access, and the right to know" (O'Hair & Heath, 2005, pp. 4–5). This understanding can provide insight to health professionals regarding the construction of successful intervention campaigns that improve health through modifying behaviors.

For better or worse, the media can play an important role in how the public obtains reliable and appropriate health information, especially that which is crisis-related. By understanding how a community might typically respond to a particular health crisis, public relations experts, medical professionals, and community leaders can take advantage of these tendencies and/or engage the CCC's help in preparation and in making their health campaigns more accessible, believable, and effective. When community members utilize the media to engage in open public dialogue with each other, they have the opportunity to help propagate important and credible health information.

Conclusion

The purpose of this chapter has been to overview the common types of biological and health-related crises and their outcomes. In addition, specific local-level strategies that could be facilitated by local CCCs were addressed. Finally, the chapter outlined the CCC's role in information collection and dissemination and the relationship it can establish and maintain with the local media to help inform and prepare the community for any health-based crises that threatens them.

Kreps et al. (2005) stressed that "communication … is the primary social process for sharing relevant information among the key policymakers, security personnel, healthcare providers, and members of the general public" (p. 191). Members of the community depend on their local leaders during times of crises. In turn, these leaders need to be able to depend on each other and be able to exchange the necessary information to coordinate their efforts. O'Boyle and Simms (2005) state, "One of today's major challenges for health leaders is to bring together the perspectives and resources of the medical care industry, the population-based public health sector, and the community to develop community-based health systems that work" (p. 251). The call for coordinated communication—a main function of the CCC—goes beyond function; it reflects a value of social justice, which provides a foundation for public health practitioners and community leaders who strive to make a difference in their community (O'Boyle & Simms, 2005). In the following section are the websites mentioned in this chapter and other helpful resources for understanding and for preparing for health-related or biological crises:

- Questions and Answers on Ebola: http://www.cdc.gov/vhf/ebola/outbreaks/2014-west-africa/united-states-imported-case.html

- Four influenza alert levels of the World Health Organization (WHO): http://www.cidrap.umn.edu/news-perspective/2013/06/who-proposes-new-pandemic-flu-alert-levels
- Centers for Disease Control and Prevention (CDC) Guide to Crisis Emergency and Risk Communication: http://emergency.cdc.gov/cerc/

References

Action, L. K., & Ledlow, G. R. (2005). Bioterrorism: Is the nation's healthcare system prepared? In J. A. Johnson, M. H. Kennedy, & N. Delener (Eds.), *Community preparedness and response to terrorism*. Vol. 2: *The role of community organizations and business* (pp. 107–23). Westport, CT: Praeger.

Ahern, J., Galea, S., Resnick, H., Kilpatrick, D., Bucuvalas, M., Gold, J., & Vlahov, D. (2002). Television images and psychological symptoms after the September 11 terrorist attacks. *Psychiatry*, 65, 289–300.

American College of Emergency Physicians. (2012). Health care system surge capacity recognition, preparedness, and response. http://www.acep.org/content.aspx?id=29506

Associated Press. (2012, April 3). Haiti sees jump in cholera cases: UN; 77 new cases reported since beginning of rainy season in March. *New York Daily News*. Retrieved from http://www.nydailynews.com/life-style/health/haiti-sees-jump-cholera-cases-77-cases-reported-beginning-rainy-season-march-article-1.1055396?localLinksEnabled=false

Barbisch, D. (2005). Regional responses to terrorism and other medical disasters: Developing sustainable surge capacity. In G. R. Ledlow, J. A. Johnson, & W. J. Jones (Eds.), *Community preparedness and response to terrorism*, Vol. 1: *The terrorist threat and community response* (pp. 77–88). Westport, CT: Praeger.

Becker, J. A. H., & Thompson, S. (2005). The intersection of terrorism, interpersonal communication, and health. In H. D. O'Hair, R. L. Heath, & G. R. Ledlow (Eds.), *Community preparedness and response to terrorism*, Vol. 3: *Communication and the media* (pp. 47–64). Westport, CT: Praeger.

Brodie, M., Flournoy, R. E., Altman, D. E., Blendon, R. J., Benson, J. M., & Rosenbaum, M. D. (2000). Health information, the Internet, and the digital divide. *Health Affairs*, 19(6), 255–265, doi:10.1377/hlthaff.19.6.255

Burke, E. (1999). *Corporate community relations: The principle of neighbor choice*. Westport, CT: Quorum Books.

Centers for Disease Control and Prevention. (2010). Summary of trial crisis and emergency risk communications (CERC) training, December 8–9, Atlanta, GA. Retrieved March 24, 2016, from http://www.cdc.gov/h1n1flu/statelocal/cerc_training.htm

Centers for Disease Control and Prevention. (2014). Cases of Ebola diagnosed in the United States. Retrieved August 2, 2015, from http://www.cdc.gov/vhf/ebola/outbreaks/2014-west-africa/united-states-imported-case.html

Centers for Disease Control and Prevention. (2015). 2014 Ebola outbreak in West Africa-Outbreak distribution map. Retrieved August 2, 2015, from http://www.cdc.gov/vhf/ebola/outbreaks/2014-west-africa/distribution-map.html#areas

Cole, L. A. (2012, January 18). Bioterrorism: Still a threat to the United States. CTC Sentinel. Retrieved August 2, 2015, from https://www.ctc.usma.edu/posts/bioterrorism-still-a-threat-to-the-united-states

Craig, M., Olaniran, B. A., Scholl, J. C., & Williams, D. E. (2006). Crisis communication in public arenas. *Public Relations Review, 32*, 171–73.

Curran, D. R. (2005). Influenza vaccination clinic: A mass vaccination exercise. In G. R. Ledlow, J. A. Johnson, & W. J. Jones (Eds.), *Community preparedness and response to terrorism*, Vol. 1: *The terrorist threat and community response* (pp. 169–82). Westport, CT: Praeger.

Danticat, E. (2012, February). Anniversary blues. *Progressive, 76*(2), 20–22.

de Perio, M. A., Wiegand, D. M., Brueck, S. E., & Scholl, J. C. (2014). Take your best shot—flu vaccinations can prevent serious illness, but many employees still forgo them. *American School Board Journal, 201*(4), 21–23.

Dezenhall, E., & Weber, J. (2007). *Damage control: Why everything you know about crisis management is wrong*. London: Penguin.

Du, J. (2004). Bioterrorism: How has it been used? What can it do? How prepared are we? Retrieved from www.isop.ucla.edu/article.asp?parentid=1352

Dzenowagis, J. (2011). Bridging the digital divide in health. *UN Chronicle, 48*(3), 27–29.

Fox, M. (2014, November 6). It's over: Texas' Ebola outbreak has ended. NBC News. Retrieved August 2, 2015, from http://www.nbcnews.com/storyline/ebola-virus-outbreak/its-over-texas-ebola-outbreak-has-ended-n242931

George, A. M., & Evuleocha, S. (2003). Denials, excuses, justifications, and apologies: Restoring tarnished reputations after the year of corporate malfeasance, what worked and what didn't. *Proceedings of the 2003 Association for Business Communication Annual Convention*, Albuquerque, New Mexico.

Gibson, E. (2010, September 30). FDA moved too slowly on J & J recall. *Bloomberg Business Week*. Retrieved from http://www.businessweek.com/news/2010-09-30/fda-moved-too-slowly-on-j-j-recall-official-says.html

Hepatitis Attorney. (2012, February 19). Lubbock's Cheddar's Casual Café costs public health $250,000 due to Hepatitis A. http://www.hepatitisblog.com/hepatitis-a-watch/

Inungu, J. N. (2005). Overview of biological agents and compounds with potential bioterrorism applications. In J. A. Johnson, M. H. Kennedy, & N. Delener (Eds.), *Community preparedness and response to terrorism*, Vol. 2: *The role of community organizations and business* (pp. 1–13). Westport, CT: Praeger.

Kamel Boulos, M. N., Resch, B., Crowley, D. N., Breslin, J. G., Sohn, G., Burtner, R., Pike, W. A., Jezierski, E., & Slayer Chuang, K. (2011). Crowdsourcing, citizen sensing and sensor web technologies for public and environmental health surveillance and crisis management: Trends, OGC standards and application examples. *International Journal of Health Geographics 10*(67). Retrieved August 3, 2015, from http://www.biomedcentral.com/content/pdf/1476-072X-10–67.pdf

Kreps, G. L., Alibek, K., Bailey, C., Neunauser, L., Rowan, K. E., & Sparks, L. (2005). The critical role of communication to prepare for biological threats: Prevention, mobilization, and response. In H. D. O'Hair, R. L. Heath, & G. R. Ledlow (Eds.), *Community preparedness and response to terrorism*, Vol. 3: *Communication and the media* (pp. 191–210). Westport, CT: Praeger.

Kulic, K. R. (2003). An account of group work with family members of 9/11. *Journal for Specialists in Group Work*, *28*, 195–98.

Leaffer, T., & Mickelberg, L. (May/June 2006). The digital health care revolution: Empower health consumers. *The Futurist*, 53–57.

Lerbinger, O. (2012). *The crisis manager: Facing disasters, conflicts, and failures* (2nd ed.). New York: Routledge.

National Terror Alert. (2012). Biological attack. http://www.nationalterroralert.com/biologicalattack/

Niederhoffer, K. G., & Pennebaker, J. W. (2002). Sharing one's story: On the benefits of writing or talking about emotional experience. In C. R. Snyder & S. J. Lopez (Eds.), *Handbook of positive psychology* (pp. 573–83). London: Oxford University.

Nightingale, S. (2010). Sexually transmitted diseases remain trouble spot for Lubbock County. http://lubbockonline.com/health/2010–08-29/sexually-transmitted-diseases-remain-trouble-spot-lubbock-county

Nöggerath, J., Geller, R. J., & Gusikov, V. K. (2011). Fukushima: The myth of safety, the reality of geoscience. *Bulletin of the Atomic Scientists*, *67*(5), 37–46, doi: 10.1177/0096340211421607

O'Boyle, I., & Simms, M. (2005). Public health ethics, business, and terrorism. In J. A. Johnson, M. H. Kennedy, & N. Delener (Eds.), *Community preparedness and response to terrorism*, Vol. 2: *The role of community organizations and business* (pp. 249–66). Westport, CT: Praeger.

O'Hair, H. D., & Heath, R. L. (2005). Conceptualizing communication and terrorism. In H. D. O'Hair, R. L. Heath, & G. R. Ledlow (Eds.), *Community preparedness and response to* terrorism, Vol. 3: *Communication and the media* (pp. 1–13). Westport, CT: Praeger.

Olaniran, B. A., Scholl, J. C., Williams, D. E., & Boyer, L. (2011). Johnson and Johnson phantom recall: A fall from grace or a re-visit of the ghost of the past. *Public Relations Review*, 38, 153–55, doi:10.1016/j.pubrev.2011.08.001.

Olaniran, B. A., & Williams, D. E. (2001). Anticipatory model of crisis management: A vigilant response to technological crises. In R. L. Heath & G. Vasquez (Eds.), *Handbook of Public Relations* (pp. 487–500). Thousand Oaks, CA: Sage.

Pennebaker, J. W. (1997). Writing about emotional experiences as a therapeutic process. *Psychological Science*, 8, 162–66.

Rosen, C. S., Greene, C. J., Young, H. E., & Norris, F. H. (2010). Tailoring disaster mental health services to diverse needs: An analysis of 36 crisis counseling projects. *Health & Social Work*, *35*(3), 211–20.

Ruffin, A. B., Cogdill, K., Kutty, L., & Hudson-Ochillo, M. (2005). Access to electronic health information for the public: Analysis of fifty-three funded projects. *Library Trends*, *53*(3), 434–53.

Saylor, C. F., Cowart, B. L., Lipovsky, J. A., Jackson, C., Finch, A. J., Jr. (2003). Media exposure to September 11. *American Behavioral Sciences*, 46, 1622–42.

Schneider, M-J. (2006). *Introduction to public health* (2nd ed.). Boston: Jones and Bartlett.
Schnirring, L. (2013, June 10). WHO proposes new pandemic flu alert levels. Center for Infectious Disease Research and Policy. Retrieved August 4, 2015, from http://www.cidrap.umn.edu/news-perspective/2013/06/who-proposes-new-pandemic-flu-alert-levels
Scholl, J. C., Olaniran, B. A., Williams, D. E., & Heuman, A. (2007). Data collection for crisis communication centers: Community preferences for terror threat and act information. In A. Merkidze (Ed.), *Terrorism issues: Threat assessment, consequences and prevention* (pp. 153–74). Hauppauge, NY: Nova Science.
Scholl, J. C., Williams, D. C., & Olaniran, B. A. (2005). Preparing for terrorism: A rationale for crisis communication center. In H. O'Hair, R. Heath, & G. Ledlow (Eds.), *Community preparedness and response to terrorism* (pp. 243–68). Westport, CT: Praeger.
Sheely, G. (2003). *Middletown America: One town's passage from trauma to hope*. New York: Random House.
Taintor, Z. (2003). Assessing mental health needs. In B. Levy & V. Sidel (Eds.), *Terrorism and public health: A balanced approach to strengthening systems and protecting people* (pp. 49–68). Oxford: Oxford University Press.
von Lubitz, D. K. J. E., & Lary, M. J. (in press). Bioterrorism, medical readiness, and distributed simulation training of first responders. In J. Johnson, M. Cwiek, & G. Ludlow (Eds.). *Community and response to terrorism* (Vol. II): *A community of organizations: Responsibilities and collaboration among all sectors of society*. Westport, CT: Praeger.
Zinn, H. (2003). *A people's history of the United States*. New York: HarperCollins.

Case Study

Work vows dod's live anthrax incident "will never happen again"

Source: By Terri Moon Cronk, "DoD News, Defense Media Activity." http://www.defense.gov/News-Article-View/Article/612686

WASHINGTON, July 23, 2015—Calling the Defense Department's inadvertent May delivery of live anthrax to a commercial lab "a massive institutional failure," Deputy Defense Secretary Bob Work today outlined findings of the review and steps DoD will take to ensure it never happens again.

But given the failure, Work said at a Pentagon news conference, he, Defense Secretary Ash Carter, and DoD senior leadership "take absolutely no comfort in this."

The live-anthrax shipment to a commercial lab was narrowed down to one of four DoD labs, he said, and added that of the total batches in the lab's inventory, more than half tested positive for live anthrax—17 of 33.

DoD ships inactivated samples of anthrax spores to outside labs for research and development of DoD countermeasures to protect U.S. troops, allies, partners, and the American public from biological attack, the deputy defense secretary said.

"[In] the past, some nations have produced biological weapons, and we cannot be certain that rogue nations or terrorist groups won't do the same in the future," he said of DoD's work with anthrax.

Four Labs Handle Anthrax: The four labs that handle anthrax are the Army's labs at Dugway Proving Ground in Utah and in Maryland, the Edgewood Chemical Biological Center, the Army Medical Research Institute of Infectious Diseases, and the Naval Medical Research Center.

When the incident unfolded, the Centers for Disease Control began onsite investigations, and Work directed Frank Kendall, the undersecretary of defense for acquisition, technology, and logistics, to conduct the 30-day review of the labs' procedures, processes, and protocols associated with inactivating anthrax spores.

The deputy secretary said he wanted to determine whether the accidental delivery was a one-time failure or a "more systemic problem" in DoD's biohazard safety procedures, and he assembled a team of experts from the Departments of Agriculture, Defense, Energy and Homeland Security, the FBI, academia, and industry.

Review Committee: The review committee found the four labs had irradiated—destroyed the bacterium with radiation—149 batches of live anthrax spores and reported them inactivated and safe for testing since 2003. All were accounted for and tested or destroyed, he said.

Other findings: 53of the 149 batches are no longer in the DoD inventory or available for testing, and recipients were told to destroy remaining batches.

—Of the 96 batches to test in DoD inventory at Dugway Proving Ground, 17 of 33 tested positive for regrowth/live anthrax.

"Obviously, when over half of anthrax batches presumed to be inactivated instead prove to contain live anthrax spores, we have a major problem, and the numbers confirm this judgment," Work said.

"We know over the past 12 years, 86 laboratories in 20 states, the District of Columbia, and seven foreign countries ultimately received what were supposedly inactivated spores that originated at Dugway," he added.

The deputy defense secretary said he expects the numbers to increase with secondary shipments, which CDC is responsible for tracking.

No Infections Resulted: Work said the live anthrax was delivered in small amounts of liquid in "extremely low concentrations," and all were handled in labs. Not a single infection resulted in lab workers in 12 years, he added.

Three main causes led to the incident. Work said:

—No national standards exist for anthrax inactivation procedures;

—Several factors at Dugway led to a higher likelihood of misidentifying the live spores; and

—Biohazard protocol is not standardized across DoD, because the labs report to multiple chains of command and two services.

"The American public expects much more from us, and we should expect much more from ourselves," Work said. "Secretary Carter has made plain he expects these issues be dealt with swiftly and comprehensively to ensure that a failure of this sort never happens again."

Work said he directed Kendall's office to:

—Work with DoD stakeholders, the CDC, and other relevant departments and agencies to develop standardized irradiation and viability testing protocols for all DOD labs that work with spore-producing organisms like anthrax;

—Ensure sufficient funding is available through the Chemical and Biological Defense Program for research into the development of standardized irradiation and viability testing protocols;

—Review and revise as necessary DoD biosafety and biosecurity policy and ensure consistent application across DoD laboratories; and

—Oversee military department and service implementation of the review committee's recommendations.

Work said he also directed the Secretaries of the Army and Navy to investigate and develop plans that cover quality assurance, peer review, and program management, and to provide quarterly updates on progress. The army secretary was designated as DoD's executive agent for the Biological Select Agent and Toxin Biosafety Program, the deputy defense secretary added. The moratorium on production, handling, testing, and shipment will continue until the report's recommendations are addressed, he said.

Notification Led to Immediate Action

Work commended the lab that notified the CDC of the live anthrax delivery.

"Because of their notification, the department was able to take immediate action, stop all shipping of inactivated anthrax, assess our procedures, processes, and protocols; and begin to institute needed changes," he said. "In doing so, they helped ensure nobody was infected by anthrax.

"This review taught us lessons we needed to learn, and identified institutional and procedural failures we need urgently to address," he continued. "We are shocked by these failures.... DoD takes full responsibility for these failures, and we are implementing changes and recommending the establishment of procedures, processes, and protocols that will prevent such a biohazard safety failure does not happen again."

Discussion Questions:

1. What are the most significant communication lessons learned from the mishandling of anthrax, as described in this case?
2. The mishandling of anthrax can potentially lead to a biological crisis. What other types of crises might result from this situation?
3. What role(s) could a CCC play in making sure such a biological threat to the community is less likely to happen again?

· 7 ·

CULTURAL ISSUES IN ANTICIPATORY CRISIS MANAGEMENT

Chapter Highlights

- A culture as defined by its worldview, ways of thinking, use of time, avoidance of uncertainty, emphasis on individual versus community goals, and the importance of hierarchies
- Susceptibility of some marginalized groups to the impacts of crisis events
- Culture and the impact on the perception of risk and on the severity of crisis events.
- The need for multifaceted approaches to adapt to the needs of the CCC's diverse audiences
- Racially Oriented Crisis Review and Response Teams (ROCRRs)

Introduction

Effectively communicating before, during, and after a crisis requires being able to send the right message to the right audience in the right way (Albrecht, 1996). More to that point, most audiences are heterogeneous enough that different messages and goals are needed to reach each segment of that audience.

Sources of crises are universal (e.g., natural disasters, health crises, and business misconduct), but the publics that affected by them are varied (Dykeman, 2005). Audiences vary in the meanings they attach to crisis events. They also react to crisis messages in different ways (Aldoory, 2009). People have their different interests and agendas, and community leaders and message designers must understand how each group or stakeholder "culture" interprets and responds to a crisis in its unique way.

Culture is often a less recognized variable in crisis communication research and management (Dykeman, 2005), but it does receive some attention in the public relations field (Olaniran & Williams, 2012): "The implication of this omission in crisis literature is that it creates a significant void in understanding how to manage multicultural issues as they influence effectiveness and efficiency of crisis management programs from both corporate and academic perspectives" (Olaniran & Williams, 2012, pp. 227–28). This void of understanding has already resulted in the poor handling of some crises in recent years, particularly the Red Cross's lack of cultural sensitivity in its response during Hurricane Katrina (Epstein, 2007). A missed opportunity in understanding diverse audiences was seen before, during, and after Hurricane Katrina hit the Gulf Coast (Veil, Mitcham, & Sellnow, 2012). The inability to meet the needs of the poor and racially disadvantaged of New Orleans during evacuation and in the hurricane's aftermath contributed to the increased death toll, and likely diminished the credibility of local, state, and national governments.

Cultural insensitivity can also be seen when those who communicate with the public about a crisis fail to protect certain segments of the population from undue scrutiny and stigma. Such is often the case when immigrants and people of color are treated with suspicion or are discriminated against because they are commonly associated with a crisis or safety threat. For instance, people visiting the United States from an African country might experience mistreatment or discrimination when the fear associated with Ebola might be running high.

Matters of diversity are important in anticipatory crisis management. Culture provides a context in which a crisis emerges and is reported to the public. Culture also plays a part in the delivery of interventions and guidance and in how the public responds. As a general rule, municipalities and organizations should take culture into account in their management plans. In many cases, however, practitioners and leaders can hold mixed feelings about cultural diversity. Oliveira (2013) reports that while many practitioners can point out specific cultural elements that need to be addressed in crisis messages, they often see these elements as difficulties that often demand too much time or that create too much controversy. Nevertheless, taking culture into consideration increases the effectiveness of any crisis response and effort to safeguard the community. Rather than worry about appearing cynical about prejudice and discrimination taking place, it is better to prevent the kinds of issues that bring about bad press, lawsuits, or even riots (Waymer, 2012).

This chapter first discusses what is meant by diversity and provides a picture of diverse cultures that are typically affected by crises. Next, the reality of multiple meanings will be explicated, particularly how this plays a role in varying perceptions of risk to the public. Subsequently, a shift in paradigms or ways of thinking about crises and culture will be presented, followed by implications and recommendations for improving crisis responses and management in relation to culture and diversity.

The Diverse Community

The United States is projected to be older and more diverse in future decades. According to the U.S. Census Bureau (2012), the nonwhite Hispanic population will become the largest single group by 2060, but no group will be in the strong majority. Hispanics and Latinos currently make up 1 in 6 Americans; by 2060, they will be 1 in 3. Additionally, people aged 65 and older will increase from 43.1 million in 2012, to 92 million in 2060, bringing the ratio from 1 in 7 to 1 in 5 people in the United States. During this same time period, the Asian population in the United States will double and African Americans will increase from 13.1% to 14.7% of the nation's population. Alaskan Natives and American Indians will more than double by 2060, from 3.9 million to 6.3 million. Native Hawaiians and Pacific Islanders will increase to nearly double, from 706,000 to 1.4 million. People who identify as multiple races or ethnicities will triple in numbers, from 7.5 million to 26.7. By 2043, it is projected that the United States will become a majority-minority nation for the first time in its history, with minorities increasing from 37 percent to 57 percent of the total population by 2060.

Despite the knowledge experts have of the demographic distribution of communities, many do not have a clear or consistent understanding of multiculturalism or diversity, or their implications (Oliveira, 2013). Cultures are not simply defined and differentiated based on language use, national origin, skin color, or religion, they are also characterized by such indicators as age, religion, and education. More to the point, it is a mistake to see race or culture as little more than variables that can be measured or categorized (Waymer, 2012). Rather, each culture is defined by one or more dimensions of variability that reflect worldviews, ways of thinking, use of time, avoidance of uncertainty, emphasis on individual versus community goals, and the importance of hierarchies (Aldoory, 2009; Hofstede, 1980; 1983; Hofstede & Bond, 1984).

When telling the story of a crisis, or explaining its causes and influences, there are different voices involved. This makes defining culture a political matter (Aldoory, 2009). There is the managerial or authoritative voice, the one that tends to get more attention and credibility. Then there are the silenced, marginalized, or competing voices; these might represent employees, victims, and community members who have little power over their environment or cannot easily access community leadership. When a crisis occurs, it is most likely that there are multiple accounts of the causes and consequences of that crisis. Waymer (2012) argues that diversity "is the manifestation of these voices. Multiple voices add value to the dialogue" (p. 293).

A dialogue about cultural diversity must include those groups more vulnerable due to their position or status relative to other groups. Vulnerable populations can be defined as those with medical, sensory, physical, or cognitive disabilities. They also include older adults, people whose primary language is not English (or that of the mainstream community), people who do not have adequate access to transportation, individuals who cannot live independently because of intellectual disabilities, or those with special dietary needs or restrictions. Individuals of lower socioeconomic status are a growing segment of the U.S. population: "People living in poverty are generally unable to gain access to emergency preparedness information because they do not have access to information technology" (Veil et al., 2012, p. 205). People who have minority status because of race or nationality are also potentially vulnerable because they tend to be unduly represented among the lower socioeconomic population segments.

Cultural minorities and marginalized groups whose primary language is not the one spoken by the mainstream public are at an even greater disadvantage. Veil et al. (2012) report that many non-English-speaking ethnic minorities belong to lower socioeconomic groups. Not only are such individuals unable to understand adequately crisis preparedness messages, but they also lack the resources to receive those messages or to respond in ways that are expected of average citizens. The lack of English is not just an economic issue. Veil et al. point out that immigrants who speak a language other than English do so because they still have strong ties to their cultures of origin. Such strong ties can serve to hinder minorities' desire and ability to venture out into mainstream society and become more fluent in English or in the language that is predominant.

One implication of culture is that it can play a role in how crises emerge and impact the community: "A cultural value modifies individual's ways of thinking, judging situations, and communicating" (Oliveira, 2013, p. 270). Olaniran and Williams (2012) outline two types of culture-related crises. The first type affects a specific culture (i.e., intracultural). An example of an intracultural crisis was the Exxon *Valdez* oil spill, which occurred within the boundaries of the United States. The second type of cultural crisis spans cultural boundaries (intercultural/cross-cultural) and is exemplified by the issue with Shell and the Nigerian Ogoni people because it involved a company headquartered in Europe whose oil-drilling adversely affected the homes and livelihoods of the local people living in Nigeria.

Potential for Prejudice and Discrimination

When communicating with diverse communities, especially those who have historically been disenfranchised and marginalized, one has to understand the tendencies for discrimination, as well as these groups' past experiences with mistreatment. Discriminating against a member of such a group means to exclude, restrict, make different, or give different treatment that leads to that person having fewer advantages than members of other groups in similar situations (Dobre & Tudor, 2012). For example, deliberately alerting some neighborhoods in a community of an approaching tornado far in advance of alerting other neighborhoods would be discriminatory because this is allowing some residents more time to plan and/or evaluate than others would have. According to Dobre and Tudor (2012), cultural discrimination can be based on nationality, race, religion, gender, sexual orientation, age, disability, or language, or it can be based on a combination of any of these dimensions of variability.

Cultural prejudice and discrimination often exist because they are tied to disparities among social classes. Social class does not just reflect economics but also property ownership, prestige among the community and peers, debt-to-income ratio, and power (Allen, 2004). This goes hand in hand with discussing what some argue is the myth of the American dream and the ways in which people can attain it. When certain groups find it difficult to attain this dream, it can be difficult to realize that these individuals might not have the same access to opportunities and necessary resources. The reality of unequal access to opportunity gets overlooked, and such groups are seen as incompetent or too lazy to take advantage of those very opportunities to which they do not have access. Waymer (2012) explains:

> This is problematic because this can lead to a belief that people overwhelmingly are to blame for their "non-success"—without taking into account the socially constructed barriers present within a given society. By refusing to acknowledge, account for, and address, in part, the political, economic, and structural barriers that physically and [psychologically] separate an organization from its myriad publics, an organization sustains a suitable environment for a simmering, smoldering crisis to brew and eventually erupt. (p. 289)

Holding and perpetuating such beliefs about the disadvantaged can discourage efforts to provide these groups with the assistance they might need to prepare for, or respond to, crises, especially when their groups might be affected more than others in the community.

The disadvantaged can also be affected when they find themselves to be the victims of a crisis. According to Dykeman (2005), a society that does not accept the victim status of those affected will likely be silent on the issue or even disparage those who have suffered the most. For example, members of a poor and predominantly minority community who are devastated by a hurricane might not get sympathy from the rest of the community or nation or might not get adequate financial assistance during recovery. Outside members might question why the victims did not evacuate before the hurricane hit, and it might be the reality that many of them did not have access to transportation or did not have the means to leave town. In other cases, "if one's own culture, or the culture in which one exists, either rejects or stigmatizes a crisis victim, the victim experiences additional injury to the provoking events" (Dykeman, 2005, p. 45).

Disparities among social classes can lead to conflict. For instance, Timothy Thomas, an unarmed black man, was shot by a Cincinnati police officer on April 7, 2001. Thomas's death was another incident in a string of police shootings in the city. The protests and riots that resulted from this latest shooting proved to be a public relations nightmare for the city. The conditions leading to such upheaval can lead to a crisis that the local government needs to respond to (Waymer, 2009, 2012): "Crises create rhetorical problems that demand that sense be made of decisions and events.... In a country in which its history is filled with socioeconomic disparity, organizations must be vigilant about identifying and planning for crises that might emerge due to social class conflicts—especially in down global economic times" (Waymer, 2012, p. 290). Given the reality of social class differences and disparities, local crisis managers should include consideration of such disparities in their anticipatory strategies.

Cultural disparities are not just the potential outcomes of a crisis; they can also be the sole reason or cause of a locally based crisis. Such crises are called racially oriented crises (ROCs), such as a riot in response to a racially charged event or a widely publicized racial slur that elicits community backlash (Williams & Olaniran, 2012). Groups who use the phrase "Black Lives Matter" to protest recent incidences of police brutality of unarmed people of color might become emotionally charged when confronted with counter messages that might be construed as racist. ROCs are particularly unique in that the group, organization, or community held responsible can pay a tremendous price for the action or language used. Williams and Olaniran (2012) make these observations:

> Attacks on individual or organizational character, stemming from ROCs, tend to generate more negative responses from the general public and stakeholders. While protests and strikes are typical outlets for groups angered by any type of crisis, ROCs join with environmental crises in generating a more significant response from a wide range of affected or angered stakeholders and segments of the public. (p. 302)

ROCs can be very difficult to manage because they are difficult to anticipate, they sometimes require responders to choose among certain strategies that can be polarizing, and they can do devastating damage to an organization or to a community. Responses that attempt to subdue the ROC often compel leaders to change certain policies in order to encourage more inclusive and less discriminatory practices or, at the very least, publicly condemn the racially charged action or language.

Coombs's (2007, 2009) Situational Crisis Communication Theory can shed light on how some racially oriented crises are handled. Coombs (2007) explains that when organizations or individuals deny the action or language, they either diminish (justify it), rebuild (apologize), or bolster (by their own reputation or position) the act. Furthermore, in responding to the racially based accusation, one can respond with a number of different strategies, such as attacking the accuser(s), diminishing the act, scapegoating, apologizing, and denying (Coombs, 2007). The strategy of denial in particular is a more assertive one, according to Coombs (2007, 2009).

It might be uncomfortable for members of the mainstream culture to include matters of diversity in pre-crisis planning. Also, it might seem more expedient to engage in denial or in diminishing strategies, such as those outlined by Coombs (2007, 2009). Many organizations avoid the issue of diversity or treat it in limited capacities because it might suggest that a sexist or racist culture indeed exists. Waymer (2012) points out that many organizations and municipalities serve, employ, and relate with members of diverse communities; therefore, issues of diversity are inescapable. Fortunately, organizations are increasingly incorporating diversity in their policies and procedures, whether it is out of goodwill or mandate. Nonetheless, this is not enough evidence to say they are adequately including diversity in their pre-crisis management plans, according to Waymer (2012).

Inadequate attention to culture when planning crisis responses invites mistreatment and the increase of vulnerability of certain minorities. Whether intentional or unintentional, prejudice and discrimination often stems from these groups being underrepresented or having few advocates that can make them more visible to the mainstream: "Because they are typically misunderstood, both verbally and socially, non-English speakers are a trapped, muted group, unlikely to know about or ask for help regarding emergency preparedness" (Veil et al., 2012, p. 216). Many minority groups are able to maintain close ties within their families, religious groups, and ethnic communities; however, these communities often lack leadership that has strong ties to city governance.

Officials and crisis managers should be aware that local organizations and industries have their own cultures, which has implications for certain factors feeding into crises. Organizational culture consists of commonly shared values, beliefs, and behaviors that are perpetuated through stories, artifacts, norms, and rituals, which often fuel destructive organizational communication patterns, such as bullying, ostracism, and discriminatory practices (Lutgen-Sandvik, Namie, & Namie, 2009). Waymer (2012) explains that through stories shared within and without the organization,

> people gain a sense of acceptable and unacceptable behaviors, which helps define the power and politics in operation; in an organizational context, the interpretation of these messages can affect employees' sense of empowerment by helping or hindering a sense of belonging and unity. (p. 292)

Crisis and culture are inextricably linked, and because an organization's culture is manifested in its everyday operations, it is unwise to prepare for a crisis without considering the cultural ramifications.

Challenges Faced by Vulnerable Populations

In some ways, culture might pose its own barrier to the effective use and reception of crisis messages. This is not to say the cultural groups are inherently problematic. The risk factors of those groups, the distrust they might feel toward certain authorities, and their lack of information and access to information sources often hinder these groups' ability to benefit from crisis messages. Varying risk environments are also a factor, especially when groups travel to a new location that has risks with which they are unfamiliar. For example, "Mexican immigrants finding employment in Colorado may have never experienced blizzard conditions" (Veil et al., 2012, p. 218), and therefore might face difficulties responding appropriately to crises if they cannot handle severe winter weather conditions. Culture can also serve as its own barrier when a language gap exists. For instance, although many messages generated by FEMA and by the Red Cross have literal translations from English to other languages, such as Spanish, these messages are often not designed or implemented to build these groups' literacy or trust, and they tend not to focus attention on cultural variations (Andrulis, Siddiqui, & Gantner, 2007).

When communicating directives or instructions that are intended to safeguard the population, one cannot assume that all segments of that population will adhere to or accept the message. For instance, Andrulis et al. (2007) claim that racial minorities tend to be more vulnerable because they hold different perceptions of risk, suffer from language and cultural barriers, have more distrust of message sources, rely more on informal sources of information (e.g., word of mouth), and tend to lack the knowledge to take proactive and protective measures. Also significant is the fact that minority populations will be less likely to accept or to attend to warning messages from authorizes, such as governmental figures, because of the distrust built up because of generations of discrimination and prejudice. Among various ethnic and disenfranchised communities there is a distrust of government officials and media outlets (Veil et al., 2012).

Prejudice, discrimination, and the distrust that ensues can contribute to diminished crisis responses by diverse members. As a result, these segments of the population are often put at greater risk for the consequences or aftereffects of crises. The emergency managers interviewed by Veil et al. (2012) suggest that many minority groups need "additional prompting or assurances from trusted sources because of distrust of government or law enforcement.... Much of the distrust can be attributed to misunderstanding of the function of government and law enforcement" (pp. 213–14). Another potential barrier to adequate responses by the community could be trying to communicate with immigrants who are undocumented, who likely fear encounters with law enforcement out of fear of being discovered and deported back to their home countries.

One barrier to ensuring the safety of vulnerable populations in times of crisis has to do with their limited ability to prepare for disasters. For example, not all households have the means or knowledge to prepare a first-aid or disaster kit. When one is living paycheck to paycheck, it might seem difficult or impossible to store three to five days of food and other rations. The emergency managers in the Veil et al. (2012) study reported that preparedness campaigns tend to be for those who can afford the materials necessary to respond to crises. From the perspective of minority and disenfranchised individuals, it is a choice between buying groceries to last the family for the week and stocking the pantry with canned goods that are reserved for extreme emergencies. Furthermore, many immigrant families struggle to procure supplies they need for *today* and are unlikely to think in terms of a possible disaster. These groups also typically do not have adequate access to the Internet, which leaves them going without crucial information or detailed crisis-related instructions, such as evacuation protocol, where to go for shelter, and what services to call for help.

More needs to be done to inform members of vulnerable populations about disaster kits and about the need to prepare extra resources. But because of economic challenges, many emergency managers see such efforts as a waste of time. Considering these factors, emergency managers might ignore or give inadequate consideration to disaster preparedness within these population segments. Such inaction, however, has proven to be disastrous, as was the case for New Orleans and for other affected areas during Hurricane Katrina. During this disaster, many disadvantaged individuals suffered high death counts, and government at all levels (local, state, and national) suffered diminished credibility (Veil et al., 2012).

Being culturally sensitive and meeting the needs of diverse communities requires leaders and crisis managers to determine which of the community's industries are more susceptible to a crisis, and which cultural groups are most affected. For instance, healthcare organizations tend to suffer from white-collar crime, and higher educational institutions tend to see sexual harassment issues (Waymer, 2012). Industrial pollution in some areas has reached crisis levels, and Douglas and Wildavsky (1983) attribute increased concern over industrial pollution and associated cancer risks with the surging environmental movement to eliminate involuntary exposure to industrial toxins. People who do not have the resources or financial means to move to safer areas within a community are often hit hardest by pollution or by the effects of natural disasters (e.g., flooding after a hurricane). Racial and ethnic minorities make up a disproportionate percentage of these groups.

Multiple Meanings and Perceptions of Risk

Communication is a process of sharing and co-creating meaning through the use of symbols (Griffin, 2006). The preference for certain types of symbols (e.g., language, jargon, and nonverbal cues) varies across culture. This variance can hinder effective meaning making and exchange from message source to intended target. In other words, cultural differences can lead to variations in the way in which groups within a community respond to important crisis-related information. For example, An, Park, Cho, and Berger (2010) examined collectivistic and individualistic students' responses to a hypothetical news story about a crisis event. They found that collectivistic individuals responded with more anger and negative impressions toward the organization being discussed than did those who exhibited more individualistic tendencies. This was the case when the message was written in a more individualistic tone. In general, cultural sensitivity in crisis communication involves "a discursive process that creates multiple meanings among both practitioners and audiences" (Curtin, 2012, p. 470).

One of the most important meanings to be conveyed is the realistic and accurate perception of the risk associated with a crisis or threat. Veil et al. (2012) report that much of the research on health campaigns includes studies of how to convey risks in an accurate way. Ideally, these campaigns are designed to resonate with audience needs, values, backgrounds, levels of health literacy, and personal and group experiences. What these efforts fail to account for are the uneven risk levels across populations, and, unfortunately, those populations with the highest levels of risk and vulnerability have the most difficult time accessing the necessary resources to learn of those risks and to combat them.

Additionally, not all cultures perceive a risk or threat with the same severity. Racial and ethnic minorities tend to use different communication channels and have different perceptions of risk, which makes them less likely than other groups to receive and to benefit from the information they need (Lachlan, Burke, Spence, & Griffin, 2009). Moreover, in some cultures, simply talking about a possible crisis is perceived to be the same as bringing it on; therefore, crucial crisis messages can be ignored. The value placed on preparedness varies across cultures; some groups are not even familiar with the notion (Veil et al., 2012). Getting diverse audiences to understand the level of risk or threat is crucial to maintaining public safety and minimizing damage from catastrophe. Tansey and Rayner (2009) explain:

> The dominant model of risk communication essentially is one of information transmission with the goal of educating the recipient to arrive at a rational understanding of the probable risks. The main concern is how to pass quantitative information about the probabilities and consequences of events from one information bearer … to another … through a medium … with the minimum of distortion. (p. 60)

Tansey and Rayner (2009) further point out that transmitting information is just one part of communicating. It also requires sharing this meaning among individuals and groups in such a way as to build, maintain, or salvage healthy relationships across the community.

Cultural theory provides a framework for how managers can manage risks for diverse community members. Cultural theory argues that risks determined and managed by the principles are held within the structures of social organization (Tansey & Rayner, 2008). The ultimate concern should be on the dynamic relationships among individuals. These relationships are formed in part by the institutions in which they are created and maintained. According to cultural theory, the organizational structure is what feeds into the perception of a risk. For example, if employees at a chemical plant are accustomed to following a standard for when a certain chemical level becomes dangerous, they might be "institutionalized" against thinking that levels approaching the danger zone should raise alarm. Employees might even discourage others from reporting such levels as dangerous because they are still within the normally accepted levels, and that it might be a hassle to lodge a report. Therefore, the culture of such an organization is such that vigilance beyond what is dictated by the company standards could subject an employee to scrutiny or sanctions by other members of that culture.

This theory is often considered a unique way to approach risk perception because it assumes that community and audience members are active participants rather than mere observers. In addition, being proactive means encouraging strategies that are beyond what might seem to be conventional in approach (Tansey & Rayner, 2010). Viewing individuals as active participants portrays them as members who understand that some risks receive attention and that others are hidden from view (Douglas, 1985). For example, a chemical plant employee notices that the toxicity level of a chemical is just below what would be considered a security breach. That employee can choose whether or not to draw attention to that potential breach. Such a decision is influenced by the pervasive organizational culture.

Curtin (2012) presents a cultural-economic model that explains how meaning is conveyed and interpreted through a cultural lens. Curtin describes five moments that occur within an interaction—regulation, production, consumption, identity, and representation—that are interactive and mutually influential. The moment of regulation involves the communication of expectations regarding what is right or wrong. These expectations can change over time and vary across groups. For instance, while a city government might maintain that pointing out the community's vulnerabilities is the right thing to do to plan for a crisis, some minority groups might see that as only bringing about trouble, especially if that kind of attention can bring about prejudice or even violence.

Messages have meanings and significance for the sender and for the receivers, and the meanings arise from cultural and organizational parameters. This embodies the moment of production. Meanings that senders convey are also shaped (and sometimes hindered) by the technology used to disseminate the messages (Curtin, 2012). For example, if a crisis message is disseminated through an organizational website, some cultural groups might not see that message as significant or worthy of attention because it was not sent through more media-rich forms, such as television, radio, or word of mouth. In this case, the choice of medium can affect the meaning interpreted and the attention paid by those who receive the message.

Production eventually leads to consumption, which refers to how targets use messages and create new meanings embedded within them. For example, regarding a message about specific instructions about a looming tornado, receivers can do one or more things with that message. One option is to follow the instructions to the letter and to take action; another is simply to pass along those instructions to others. A third option is to see those instructions not as directives but as choices; the instructions are there for the receivers to take or to leave.

Identity is yet another moment in the interaction. Identity is central to the model, according to Curtin (2012), because our social networks and cultural groups influence how we send and receive information, respond to messages, and understand crises as they happen. Identities are complex and multifaceted, just as groups are. For example, not all Latinos living in a particular region are the same; some are first-generation Mexicans, while others are second- or third-generation individuals who claim a Spanish (not Mexican) lineage. Some Latinos might be religious and maintain close ties with their church communities; others are secular and might not respond well to a message with spiritual overtones. Message designers have to consider this complexity when trying to design messages for these communities.

Finally, moments of representation include "the materials produced and the meanings encoded in them, the ideas they represent" (Curtin, 2012, p. 471). Representations include how a message is transmitted, the structure of that message, and its content. Curtin maintains that these five moments are mutually exclusive, and that they often overlap, which sets in motion a process of creating, sharing, and negotiating meanings within communities, and among community members, municipal leaders, and crisis managers. It is through the interplay of these moments that crises can give rise to new meanings and connections among community members. Curtin (2012) explains:

> While corporations are often empowered in globalized situations through their control [of] resources, this perspective also acknowledges that situational specifics can lead to varying degrees of empowerment. For example, many apparel manufacturers have had to adopt more socially responsible manufacturing processes in response to consumer demands. Power is always relative, then, and varies according to each relationship. (p. 472)

Paradigm Shift: From Traditional to Multifaceted Approaches

Taking an anticipatory crisis communication stance with diverse audiences requires a change in the way in which experts and scholars research and apply crisis communication. Practitioners have to consider the audience as consisting of marginalized and disenfranchised groups who might not reap the same benefits of the crisis information they receive. Moreover, a shift in thinking requires debunking the notion that "there typically is not time to tailor messages to diverse publics" (Veil et al., 2012, p. 204).

More traditional ways of viewing crises and cultures have focused on defining characteristics of African American and Latino cultures and how they respond to risk and crisis messages (Aldoory, 2009). Efforts have also focused on creating Spanish language equivalents of messages that are usually disseminated to English-speaking audiences, overlooking the fact that literal translations are often inadequate and inaccurate. In addition, sex and gender have been discussed in terms of how these constructs are social in how they are learned and enacted in social life (Beck, 2001).

In addition to the study of ethnic groups and their characteristics, other traditional approaches have involved individualized interventions, meaning that the purpose of crisis management is to help the individual change to fit the overall system (Dykeman, 2005). This assumption stems from the notion that people share a widely accepted system of normal behaviors and expectations, regardless of culture or background. Also, it has traditionally been assumed that a standard, formal intervention is preferable to that which is upheld by the natural system (i.e., that of a particular culture). Perhaps the most shortsighted assumption is that the crisis manager is typically aware of his or her own assumptions about culture.

To move beyond traditional perspectives means moving past simplistic cultural boundaries as determinants of behavior. One way to break the mold is to acknowledge "the great diversity among members of the same ethnic group or gender, which has complicated the practice of risk communication" (Aldoory, 2009, p. 234). This involves considering religiosity, disability, socioeconomic status, and age as important cultural dimensions of variability. The intersection of all of these cultural variations allows scholars and crisis managers to take a more ecological view of crisis communication and response. This perspective takes into account impacts on groups and communities rather than just on individuals and their cognitive and demographic characteristics. Progressing from individual characteristics to group and community dynamics still might not be enough to change overall thinking; such progression involves entire paradigm shifts.

Solid Society to Liquid Society

One type of paradigm shift has to do with how we view society—solid or fluid. Bauman (2000) argues that society is moving away from hard structures to fluid, software-based ones. More specifically, he argues that society has reached a remote and unreachable global state that has moved away from a connectedness that used to characterize the human experience. As it pertains to crisis communication and management, a solid society assumes that there are clear borders separating cultures, centralized systems of leadership and governance, and a system of mass communication that can reach the largest audience possible (Falkheimer & Heide, 2012).

On the other hand, a liquid society is a highly complex system whose members participate at various levels on behalf of multiple groups, and different groups get to participate (i.e., opposite of centralization). While a solid society assumes that order is the norm, a liquid society is characterized by "insecurity, fragmentation, individualism, and increased reflexivity—concepts with clear connections to and great relevance for crisis communication" (Falkheimer & Heide, 2012, p. 38). Viewing society as liquid assumes that borders between regions and countries are not solid enough to contain the effects of certain events and disasters. In fact, what happens in one part of the world can reach other areas. For instance, Oliveira (2013) reminds us that the subprime mortgage crisis that hit the United States made an impact on other countries, and proved to be more than just a U.S. crisis.

Contrary to Bauman's (2000) arguably bleak view that society has become more insecure, such a paradigm shift might serve crisis managers in the sense that more diverse voices might be involved in shaping the meaning of and response to crisis events. This shift also could allow for increased adaptability in how decisions are made. Realizing that society has become more interconnected and globalized is important to crisis management and communication, especially when modern technology and social media can turn a local crisis into an international event in mere minutes (Oliveira, 2013).

Sender Orientation to Participant Orientation

Falkheimer and Heide (2012) also advocate changing the perspective of crisis communication from sender orientation to participant orientation. They point out that in a culture with a participant orientation, targets and consumers of messages use various media forms to convey their views. Social media venues like Facebook and LinkedIn are examples of how ordinary citizens post their various issues and thus set an agenda. Moving toward a more participatory model means understanding that merely making messages available does not mean that all groups will receive and understand them the same way. Being anticipatory requires a participant orientation that emphasizes a more culture-oriented perspective, which "entails that organizations adopt a holistic perspective regarding crisis communication, which necessitates the challenging of intra-organizational, economic, social, and cultural boundaries" (Falkheimer & Heide, 2012, p. 39). According to this perspective, cultural members have the ability to question what they see and experience in their communities and in the mainstream, thus possibly changing the environment in such a way as to make it more responsive to unforeseeable events. For example, if people in poor neighborhoods have the opportunity to voice their concerns about evacuation plans (e.g., in the event of a hurricane) that do not take into account some people's lack of personal transportation, such accounts could prompt crisis managers to alter their plans to accommodate neighborhoods with those limited resources.

A shift to a more participatory culture is not without challenges. Not all citizens have equal access to media to send their own messages. Some cultures, because of their relative privilege, have more influence than others (Falkheimer & Heide, 2012). In addition, current assumptions about who gets to participate and who simply receives symbolic messages are very difficult to change. There is a need to allow multiple voices to weigh in on risks and threats to the community, which could encourage more participation. It might also be beneficial to create messages, policies, and infrastructures that are culturally sensitive and emerge out of community dialogue, not just an assumed authority from an institution that might not have a complete understanding of the cultures represented within a community (Heath & Palenchar, 2009).

Again, social media can serve as a sphere for such dialogic activity, such as the creation and promotion of a CCC Facebook page on which any community member can post comments. Palenchar and Freberg (2012) claim that "these activities are more likely to be successful when they increase the control that key publics have over the technical information, risk assessment process, and regulation of any threatening product as well as its manufacture and transportation" (p. 150). Social media might become a matter of necessity where community members can stay informed and participate, especially since many government and local agencies vary in their perception of responsibility for public information and alertness. Some managers see crisis preparedness as a two-way street. While some government officials believe in taking responsibility for providing their information, they might assert that it is the responsibility of the ethnic community to communicate their needs (Veil et al., 2012). Of course, this is assuming that members of a community can speak competently in the language of the local government, they will be taken seriously, and they have not experienced abuse or mistreatment that would reduce their trust of the government.

Reaching and Serving Multiple Audiences: Theoretical Implications

Having discussed conceptualizations of culture and diversity, explored perceptions and multiple meanings, and considered possible paradigm shifts, this chapter concludes with theoretical and practical implications for increasing the consideration and application of diversity during crisis management. As a theoretical implication, practitioners and scholars can work more to incorporate a cultural competence framework (Sue, 1991, 2001), which suggests identifying cultural expectations and documented cultural differences, and encouraging adaptive behaviors.

Another theory-based recommendation is to apply Coombs's (2009) Situational Crisis Communication Theory (SCCT), which argues for a dual approach. One point represents a more scientific or technical orientation which involves looking at phenomena in order to explain, predict, and control events. The other component of SCCT is more rhetorical and interpretative, and is concerned with how members and affected individuals make sense of what is happening. Such a varied approach accounts for diversity in accounts, explanations, and interpretations of crisis events, and can give voice to those affected who otherwise are not powerful enough to have their interests considered. This theory also allows a diversity of methodological approaches used in finding the causes of the crisis and in collecting other important data that the CCC and other entities can use.

Applied Implications and Recommendations

Regardless of the audience or cultural group being reached, perhaps the most important recommendation in disseminating crisis information is to do so quickly. Being immediate and quick often requires asking questions and getting answers from the target audience, which might be seen as an intrusion on privacy (Dykeman, 2005). To address questions of privacy, crisis responders must find ways to convey to the audience the need for such information gathering. Dykeman (2005) asserts, "In this manner, questioning may represent the only expedient manner by which differences in language, customs and behaviors can be addressed and clarified in a timely and expedient manner" (p. 47).

To reach vulnerable populations in times of crisis, one must be careful about the communication strategies used. Many emergency managers make crisis-related messages available in a variety of languages, and they tend to disseminate them on the Internet (Veil et al., 2012), which can be problematic given that many minority populations do not have the ready access to computers that community leaders might assume they have. When disseminating messages to non-English-speaking populations, alternatives to Internet dissemination should include radio and/or television in the form of public service announcements (Oliveira, 2013). Radio and television stations that serve language or ethnic segments are effective message sources (Veil et al., 2012). Reaching vulnerable audiences also means changing one's tactics, such as speaking to a town hall meeting rather than appearing on an evening television news program.

Addressing the problem of technological inaccessibility means finding ways to get crucial information to community leaders of minority and ethnic communities (Dykeman, 2005; Oliveira, 2013): "Minority group leaders such as local representatives or church leaders may play a role in the dissemination of information to their constituents" (Veil et al., 2012, p. 212). Translation from English to another language might not always have the precision required, and these community leaders can help with that translation process. In addition, these community members often know best how to reach vulnerable members, and already have their trust and respect.

Many local communities do not have the adequate resources to serve their minority populations, partly because of the lack of language translation services or training in bilingual or multilingual preparedness. For this reason, local emergency managers should have regional, state, and national centers or resources on hand to which they can refer for further guidance (Veil et al., 2012). These extra-community resources, which can be coordinated by local CCCs, can fill in the gaps in terms of translation services as well as provide intercultural preparedness training, informational materials, and guidance for resettlement or relocation. One such resource is the Office of Minority Health (OMH), a division of the Department of Health and Human Services. The OMH website (http://minorityhealth.hss.gov) provides training tools for cultural competency, information about cultural and language disparities, statistics on minority populations, online bilingual glossaries, and resources for seeking grant funding for projects. The CCC should have this and other such resources available to experts and to community members who need them. Reliance on such outside materials can help emergency managers reach underprivileged audiences more effectively, save time and money in program implementation, and thus avoid dangerously draining their own resources.

One issue or problem with translation is the assumption that a simple translation, such as from English to Spanish, is sufficient in getting the point across. Veil et al. (2012) explain the problem with this assumption:

> First, the translation process frequently fails to take into account the more complex cultural distinctions of many underrepresented populations. The decision-making process and hierarchy of influence, for example, differ widely among cultures. Expecting a message designed for the typical American citizen with a reasonable income to translate clearly to underrepresented populations is excessively optimistic at best and culturally insensitive at worst. (pp. 218–19)

Accurate and culturally sensitive translation takes into account not just literal translations of words but also the nuances of those languages, ambiguous meanings contained in many words, and how those words are understood in various contexts and situations.

Face-to-face communication, whenever possible, is necessary in order to build rapport with the local community (Veil et al., 2012). Translators working with emergency responders should be members of the community being served, especially if they have an official job or title with the local government. Local officials and emergency planners can establish relationships with faith and community groups before a crisis happens, and to organize with leaders of ethnic and minority groups to build trust and respect. Instead of being faced with the distrust that is often reserved for government officials, translators who are known for their connection to the community, might help bridge that gap and even help garner more trust and respect for the local authority structure. The CCC can play a part by providing a venue for initial meetings to take place between officials and local group representatives, as well as to maintain contact information of these groups on file that can be accessed. Of course, privacy needs to be maintained. Contact information might be limited to e-mail addresses or to ways to contact these individuals that would not involve disclosure of personal addresses or related information.

Another way to enhance translation services is to involve local law enforcement and first responders. The emergency managers, whom Veil et al. (2012) interviewed, report that many firefighters and police officers have taken classes in Spanish to relate in a more meaningful way to the people with whom they deal on a regular basis. Translation services might also be enhanced by utilizing university or college resources wherever possible. Faculty, staff, and students who demonstrate proficiency in the necessary languages can volunteer to be on call, especially in communities that might not have the funds to pay for such services. In particular, emergency managers might collaborate with language departments to set up internships or systems in which students earn credit toward their degrees for participating. Again, the CCC can play a role in identifying and in coordinating these resources.

Connections made with minority groups through their leaders can lead to more effective and efficient information dissemination in times of crises, especially if members of such communities can serve as the disseminators. Veil et al. (2012) write about a community's efforts to engage vulnerable populations through the establishment of Community Emergency Response Teams (CERTs). CERT members are trained not just in spreading key information but also in participating in search and rescue operations, first aid, victim triage, distribution of emergency kits, and other important tasks. Because cultural groups can be a source of comfort and help to those in need (Dykeman, 2005), taking advantage of the close ties that exist within cultural groups can be beneficial.

Although self-preparation is an ideal goal in educating and preparing the public for a disaster, emergency kits might be distributed to those communities that have the greatest need. Those critical of this tactic might argue that this contradicts the notion of self-preparation and self-protection, a standard recommendation in crisis planning. However, such a decision helps those individuals who are unable to follow instructions on how to compile disaster kits, even if they do possess the means.

Connections with community and ethnic group leaders, having in place appropriate and accurate translation services and employing proper communication channels, can set the stage for hands-on crisis interventions. In advocating a more general approach to culturally sensitive crisis management, Dykeman (2005) asserts that intervention can be done at four levels that are implemented in a quick and timely manner. The first level involves creating an appropriate environment (which is often a safe physical location) that assures victims a trustworthy setting in which to express emotions, such as fear or anger. The CCC can provide that safe environment where individuals who serve victims can build rapport to allow trust to develop.

The second level of intervention entails an assessment of the victim's symptoms, which could be physical ailments, psychological stress, or any other markers of crisis. These symptoms might be end products of the crisis or due to another cause—the second level requires one to figure this out. One has to be careful not to universalize symptoms to the extent that variations in symptoms are not considered. As Dykeman (2005) points out, a victim's individual responses can be due to that person's individual characteristics or to his or her lack of available resources.

At the third level, the victim needs to receive an explanation of the intervention being provided, and language translation might be an issue. The victim must be reassured that the goal of the intervention is to return her or him to pre-crisis functioning. There are four questions responders need to answer:

- What is the nature of the crisis or disaster?
- What needs to be done to or for the victim (i.e., intervention)?
- Why is it important for the victim to receive this intervention?
- What does the victim need to do on her or his part?
- Normal, pre-crisis functioning involves home, work, family, or educational aspects of the person's life. What is normal for the intervener might not be so for the victim, so it is always important to ask and verify.

At the fourth level, the intervener needs to provide information about, or referrals to, resources in the community (e.g., temporary housing, medical care, and child care). This type of follow-up intervention is extremely important because it helps the victim feel that he or she will not be abandoned. This level should also provide a way for the victim to follow up in case of questions or occurrence of repeated or new symptoms.

Racially Oriented Crisis Review and Response Teams (ROCRR)

CCCs can help organizations and individuals prepare for ROCs and be vigilant concerning potential factors. Williams and Olaniran (2012) recommend four steps: "1. Name an ROC review and response team (ROCRR) or individual. 2. Review the organization's relational history with stakeholders in terms of racial issues. 3. Evaluate potential internal triggers and external contributing factors. 4. Assess organizational readiness for an ROC" (p. 305).

These steps can be put in place within virtually any organization with the assistance of CCC personnel. It is important that the ROCRR team or individual be identified first because they are the ones who spearhead the subsequent steps. The number of team members will depend on the size of the organization, and these individuals "will need to include an individual or individuals with basic competencies in internal (organizational) and external (industry) research; rhetorical sensitivity; training program development, implementation, and evaluation; and a working knowledge of crisis management and response strategies" (p. 305). This first step might seem daunting, especially for an organization with little experience in crisis management. This is why CCC personnel can be consulted to assist in selecting and/or training this individual or team for the tasks ahead. Having a public relations expert on the team will be helpful.

The second step is reviewing the organization's history regarding race relationships with stakeholders. The ROCRR must ascertain when, how, and why race has been a part of conflict within the organization. In addition, an external assessment "will identify what media and stakeholders will discover in the event of a new crisis involving race" (Williams & Olaniran, 2012, p. 305) and other organizations that tend to be involved with the entity in question. Looking at these related organizations can shed light on what media outlets are most desirable, or ones to avoid. For example, if a local grocery chain has suffered complaints from minority customers about biased treatment, that chain might look to competing grocery stores to see what problems they have had and how they have utilized the media to handle their problems.

The third step is identifying internal and external triggers of ROCs. "The process will identify the potential for language and action triggers in the internal and external environments" (Williams & Olaniran, 2012, p. 306), such as company spokespeople who are publicly outspoken about controversial issues or about the lack of safe working conditions when employees are predominantly of minority status. More specifically, internal triggers are those that happen within the organization, such as certain hiring or firing practices and internal interactions. Internal triggers are easier to anticipate and to control than are external triggers, which might include actions of the local media, responses of customers, or industry partners.

Assessing organizational readiness for ROCs (the fourth step) depends on the accuracy and thoroughness of the previous steps. Reviewers must follow up on the relational history assessment, examine the potential internal triggers, and identify the external triggers, but it might not be possible to detect them all. Organizational readiness is about assessing the organization's vulnerabilities and should involve these questions:

- What comprises the organization's history of internal and external ROCs, and what is their potential of reoccurrence?
- What have been the primary triggers and contributing factors in previous organizational and industry ROCs?
- Who were the primary stakeholders and responding groups involved in previous organizational and in industry ROCs?
- What is the current relational status of the organization with the affected stakeholders and groups?

Reviewers should be able to answer these questions with information gathered in the previous three steps.

Implementing these actions can prove to be difficult if personnel are not properly trained or adequately experienced. Oliveira (2013) strongly endorses training initiatives that promote cultural competence. Training can include modules that encourage trainees to examine their own cultural biases, to build skills that enable them to adapt to different cultural expectations, and to expose them to diverse worldviews and perspectives. A CCC can take the lead in scheduling training sessions and in employing trainers who have backgrounds and experience in diversity and crisis management. Trainees would include spokespersons and decision makers from all segments of the community (e.g., health care, business, and first response); public relations experts charged with message design and dissemination; individuals from the media; and leaders of local communities and minority groups, particularly those from vulnerable segments. This training should be part of a larger endeavor to improve the management of diversity issues before, during, and after a crisis. Such an endeavor would supplement the training with continued briefings and updates (as community demographics change and as more information becomes available) and evaluation periods to assess managers' levels of preparation as well as decisions made.

Conclusion

Cultural diversity presents a significant impact on crisis management and communication. The purpose of this chapter was to provide a rationale for considering culture and diversity in any anticipatory crisis management plan and to gain a deeper understanding of how cultures and crises mutually influence each other. This chapter explored the varying dimensions of cultural variability and provided a picture of the diverse community, particularly with regard to prejudice and discrimination, as well as challenges faced by vulnerable populations. Next, the chapter described how meanings attached to a crisis can vary depending on the cultural group responding to it and, more specifically, explored perceptions of risks among vulnerable populaces. In addition, scholars present arguments for considering paradigm shifts in thinking about crises, particularly the shift from a solid society to a liquid society, and from a sender orientation to that which is more participant focused. Finally, the chapter presented theoretical and practical implications and recommendations for experts and practitioners as they consider diversity in their crisis management strategies.

Consult these resources for more information and tools on cultural diversity related to crisis management:

- National Association of School Psychologists (NASP): Cultural perspectives on trauma and critical response: http://www.nasponline.org/resources/crisis_safety/neat_cultural.aspx
- US Department of Health and Human Services: Cultural and Linguistic Competency for Disaster Preparedness Planning and Crisis Response: http://www.phe.gov/Preparedness/planning/abc/Pages/linguistic.aspx

References

Albrecht, S. (1996). *Crisis management for corporate self-defense*. New York: American Management Association.

Aldoory, L. (2009). The ecological perspective and other ways to (re)consider cultural factors in risk communication. In R. L. Heath & H. D. O'Hair (Eds.), *Handbook of risk and crisis communication* (pp. 227–46). New York: Routledge.

Allen, B. J. (2004). *Difference matters: Communicating social identity*. Long Grove, IL: Waveland.

An, S., Park, D., Cho, S., & Berger, B. (2010). A cross-cultural study of effective organizational crisis response strategy in the United States and South Korea. *International Journal of Strategic Communication*, 4, 225–43, doi:10.1080/1553118X.2010.515543.

Andrulis, D. P., Siddiqui, N. J., & Gantner, J. L. (2007). Preparing racially and ethnically diverse communities for public health emergencies. *Health Affairs*, 26(5), 1269–79.

Bauman, Z. (2000). *Liquid modernity*. Malden, MA: Blackwell.

Beck, C. S. (2001). *Communicating for better health: A guide through the medical mazes*. Boston: Allyn and Bacon.

Coombs, W. T. (2007). *Ongoing crisis communication* (2nd ed.). Thousand Oaks, CA: Sage.

Coombs, W. T. (2009). Crisis, crisis communication, reputation, and rhetoric. In R. L. Heath, E. L. Toth, & D. Waymer (Eds.), *Rhetorical and critical approaches to public relations*, Vol. 2 (pp. 237–52). New York: Routledge.

Curtin, P. A. (2012). Negotiating global citizenship: Mattel's 2007 recall crisis. In W. T. Coombs & S. J. Holladay (Eds.), *The handbook of crisis communication* (pp. 467–88). West Sussex, UK: Wiley-Blackwell.

Dobre, M. H., & Tudor, E. (2012). The analysis of the ethnical discrimination on the manpower market under the economic crisis. *Theoretical and Applied Economics*, *19*(9), 59–74.

Douglas, M. (1985). *Risk acceptability according to the social sciences*. New York: Russell Sage Foundation.

Douglas, M., & Wildavsky, A. (1983). *Risk and culture: An essay on the selection of technological and environmental dangers*. Berkeley: University of California Press.

Dykeman, B. F. (2005). Cultural implications of crisis intervention. *Journal of Instructional Psychology*, *32*(1), 45–58.

Epstein, K. C. (2007). Case study: American Red Cross. *Contribute*. Retrieved September 10, 2013, from http://www.contributemedia.com/trends_details.php?id=107

Falkheimer, J., & Heide, M. (2012). Participatory pre-crisis and crisis communication: A conceptual approach. In B. A. Olaniran, D. E. Williams, & W. T. Coombs (Eds.), *Pre-crisis planning, communication, and management* (pp. 37–56). New York: Peter Lang.

Griffin, E. (2006). *A first look at communication theory* (6th ed.). Boston: McGraw-Hill.

Heath, R. L., & Palenchar, M. J. (2009). *Strategic issues management: Organizations and public policy challenges* (2nd ed.). Thousand Oaks, CA: Sage.

Hofstede, G. (1980). *Culture's consequences*. Thousand Oaks, CA: Sage.

Hofstede, G. (1983). Dimensions of national cultures in fifty countries and three regions. In J. Deregkowski, S. Dziurawiec, & R. Annis (Eds.), *Explorations in cross-cultural psychology* (pp. 335–55). Lisse, Netherlands: Swets & Zeitlinger.

Hofstede, G., & Bond, M. (1984). Hofstede's culture dimensions: An independent validation using Rokeach's value survey. *Journal of Cross-Cultural Psychology*, *15*, 417–33.

Lachlan, K. A., Burke, J., Spence, P. R., & Griffin, D. (2009). Risk perceptions, race, and Hurricane Katrina. *The Howard Journal of Communications*, *20*, 295–309.

Lutgen-Sandvik, P., Namie, G., & Namie, R. (2009). Workplace bullying: Causes, consequences, and corrections. In P. Lutgen-Sandvik & B. D. Sypher (Eds.), *Destructive organizational communication: Processes, consequences, and constructive ways of organizing* (pp. 27–52). New York: Routledge.

Olaniran, B. A., & Williams, D. E. (2012). A multicultural perspective on crisis management: Shell and the Ogoni people of Nigeria. In B. A. Olaniran, D. E. Williams, & W. T. Coombs (Eds.), *Pre-crisis planning, communication, and management* (pp. 227–48). New York: Peter Lang.

Olaniran, B. A., & Williams, D. E. (2012). The need for an anticipatory perspective in crisis communication. In B. A. Olaniran, D. E. Williams, & W. T. Coombs (Eds.), *Pre-crisis planning, communication, and management* (pp. 3–15). New York: Peter Lang.

Oliveira, M. (2013). Multicultural environments and their challenges to crisis communication. *International Journal of Business Communication*, *50*(3), 253–77, doi: 10.1177/0021943613487070

Palenchar, M. J., & Freberg, K. (2012). Emergency management planning: Risk, crises, issues, and social media. In B. A. Olaniran, D. E. Williams, & W. T. Coombs (Eds.), *Pre-crisis planning, communication, and management* (pp. 147–69). New York: Peter Lang.

Sue, D. W. (1991). A model for cultural diversity training. *Journal of Counseling & Development, 70*, 99–105.

Sue, D. W. (2001). Multidimensional facets of cultural competence. *The Counseling Psychologist, 29*, 790–821.

Tansey, J., & Rayner, S. (2009). Cultural theory and risk. In R. L. Heath & H. D. O'Hair (Eds.), *Handbook of risk and crisis communication* (pp. 53–79). New York: Routledge.

United States Census Bureau (2012). U.S. Census Bureau projections show a slower growing, older, more diverse nation a half century from now. Retrieved September 10, 2013, from http://www.census.gov/newsroom/releases/archives/population/cb12-243.html

Veil, S. R., Mitcham, C., & Sellnow, T. L. (2012). Lost in translation: How emergency managers communicate preparedness with vulnerable populations. In B. A. Olaniran, D. E. Williams, & W. T. Coombs (Eds.), *Pre-crisis planning, communication, and management* (pp. 201–25). New York: Peter Lang.

Waymer, D. (2009). Liberty and justice for all? The paradox of governmental rhetoric. *Communication Quarterly, 57*, 334–51.

Waymer, D. (2012). Crisis management and communication: Pre-crisis preparation that is sensitive to diverse populations. In B. A. Olaniran, D. E. Williams, & W. T. Coombs (Eds.), *Pre-crisis planning, communication, and management* (pp. 281–98). New York: Peter Lang.

Williams, D. E., & Olaniran, B. A. (2012). Racially oriented crises: A unique challenge to crisis preparation. In B. A. Olaniran, D. E. Williams, & W. T. Coombs (Eds.), *Pre-crisis planning, communication, and management* (pp. 299–311). New York: Peter Lang.

Case Study

Identifying Community-Specific Communication Issues: Why Traditional Messages Don't Always Work

Adapted from *Effective Communication: Independent Study* (FEMA). https://training.fema.gov/emiweb/downloads/is242.pdf

Your community has a large population of immigrants and their descendants from India. As part of a community preparedness campaign, you have been asked to speak at a town meeting about how to be prepared for extreme temperatures. On the evening of your talk, you arrive early to check out the audience and notice that a large percentage is Indian. As you begin your talk, you notice that the Indian members of the audience are slowly shaking their heads back and forth sideways. You don't understand why they don't understand your message because the others in the audience are obviously engaged.

What happened during your talk? Were the Indian members of the audience not listening? Did they not understand? Did they disagree? Actually, in some parts of India, slowly shaking one's head back and forth sideways does not always mean "no." Rather, it signals, "I'm listening." The Indian members of your audience were paying every bit as much attention to your message as others.

We in the United States have strong cultural traditions based on the influence of the Western Europeans who originally established our government. Whether we realize it or not—and we often don't—these traditions shape the way we interact with other people. As our country has evolved, however, our population has experienced a huge influx of immigrants who have brought their languages, cultures, and traditions to the United States. Today, the United States hardly resembles the country that the Founding Fathers envisioned more than two centuries ago. We are older and more diverse. We are more accepting of others whose languages, cultures, and traditions are different from ours. And we recognize the inherent value of all U.S. citizens and residents, regardless of culture, age, or disability. As you saw from the case study, cultural differences reflect internal beliefs and thought patterns that cause people to react differently to the same situation. Cultural issues aren't your only considerations when communicating. Differences in age and sex, the presence of a disabling condition—and even the part of the country you live in—can affect how you communicate. To a large extent, the misunderstandings that occur involving people from different cultures have nothing to do with what they said—it's how they said it, what they did when they said it, or even who they said it to.

Discussion Questions:

1. What ethnic communities/cultural groups are represented in your community? Where are these groups located?
2. Does your community include large percentages of non-English speakers? Elderly? Disabled? Other special-needs groups? Where are these groups located?
3. What are the implications for these groups in communication during public information? Emergency communication?
4. Where can you find additional information about each of these groups that can help you plan your communication with them?

· 8 ·

ETHICAL CONSIDERATIONS IN ANTICIPATORY CRISIS MANAGEMENT

Chapter Highlights

- Deontological and utilitarian ethical approaches and how they inform crisis communication
- Local CCCs and their application of bioethics, geoethics, business ethics, and public service ethics
- The application of ethics to issues management
- The Ethical Conduct Audit (ECA) and other ethical analysis tools
- Framework for a CCC ethical code of conduct and the CCC ethics division

Introduction

In 1984, Bhopal, India, home to more than 900,000 people, experienced one of the world's worst industrial calamities. Union Carbide, now a subsidiary of Dow Chemical Company, had a chemical plant located near the city. On December 3, 1984, the plant leaked toxic gas throughout the region, which was breathed by thousands of people living there. The specific cause of the leak was the escape of methyl isocyanate gas (MIC) from a valve in the plant's underground storage tank that ruptured under compression. This crisis led to more than 23,000 deaths overall (BBC News, 2005), and about 50,000 victims needed medical treatment for facial swelling and breathing difficulties. To add to the human death toll, the city's streets were strewn with the remains of dead dogs, cats, birds, and cows.

The plant was closed after the disaster first hit. The Indian government ordered investigations, which revealed that one of the tanks storing the lethal MIC had ruptured. Even though Union Carbide settled out of court for £470 million (over $664 million), the Indian government was not satisfied. Arguing that the settlement was too low, the government sought to extradite former boss Warren Anderson for "culpable homicide." They accused Anderson of approving cost-cutting measures that resulted in compromised safety standards.

Predicting an ethics-related crisis event is not an exact science, but most businesses can reasonably expect up to 20 major ethical or misconduct disasters each year (Taub, 2002). Virtually any division of a company's operations can produce ethical scandals, including harassment, discrimination, undisclosed conflicts of interest, illegal or criminal actions, financial transgressions, consumer fraud, bribery, or regulatory abuses (Chandler, 2005). Failure to instill or to enforce an ethical code of conduct, or overlooking the ethical aspects of policies and procedures (e.g., the adherence to the safety standards at the Bhopal plant), can produce conditions that create crises. Moreover, when crises do occur, especially those that result from ethical misconduct, many organizations that either caused a crisis event or failed to prevent one often evade their ethical responsibility by blaming others or denying any wrongdoing (Olaniran & Williams, 2001).

Ethical best practices suggest that organizations, local governments, and companies be transparent and morally accountable to the publics they serve. However, it is unreasonable to assume that everyone—from entry-level employee to high-ranking official—will understand and commit to acting ethically in every situation. Even individuals with the best intentions can be influenced by social pressure, temptations, or rewards to alter their ethical decision making (Chandler, 2012).

Crisis response and management plans need to address ethical issues and reflect sound ethical policy. Well-conceptualized plans are anticipatory; they help organizations engage in pre-crisis management (Bowen, 2012). To understand the role of ethics in crisis management and to see it played out in the CCC, this chapter first reviews various perspectives of applied ethics from an anticipatory management framework. Second, different ethical issues will be applied to the AMCM, and some models for ethical crisis management will be presented. Finally, ethics-based standards and policies that are part of the CCC are introduced.

Applied Ethics and Anticipatory Crisis Management

Several explanations and definitions of a crisis have previously been given. But some additional clarification might be needed to know what constitutes an ethical crisis or a misconduct disaster. Chandler (2012) explains that "ethical misconduct disasters are specific, unexpected, and non-routine unethical events or series of unethical events that create significant operational disruptions and threaten—or are perceived to threaten—an organization's continuity of operations" (pp. 83–84). Such disasters happen because ethics violations are increasingly becoming more difficult to anticipate and manage, especially given more complex work environments.

Ethics fits well into the preplanning stages of crisis management, which would allow ethical considerations to flow through the successive phases, thus preventing many crises from occurring or resulting in significant damage. Furthermore, "Conducting analyses of responsibility can allow an organization to respond ethically to competing duties in the crisis phase and to evaluate that a moral intention was maintained and perceived by affected publics post-crisis" (Bowen, 2012, p. 58).

What makes an organization ethical? Bowen (2012) argues that ethical organizational conduct involves considering *what* and *how* information should be communicated. These considerations are part of what helps an organization act in a morally responsible manner. Responses and decisions can be *deontological* (i.e., Kant's categorical imperative; Kant, 1785/1964) or consequential (i.e., utilitarian). Although these two ethical systems are often presented as opposed to each other, anyone would be hard-pressed to find a person, policy group, or organization that adheres strictly to one or the other exclusively.

Utilitarianism focuses on the anticipated outcomes as the driving factor in decision making (Sinnott-Armstrong, 2012). Proposed by the classical thinkers Jeremy Bentham, John Stuart Mill, and Henry Sidgwick, *consequentialism* argues that the moral right or wrong behind an act depends on its consequences, as opposed to the nature of the act itself or the circumstances leading up to it. Utilitarian theory has been refined to advocate the maximization of positive outcomes and minimization of negative ones: "In other words, the greatest good for the greatest number among publics, while reducing negative impact, is the ethical course of action that serves the public interest. Harm should be minimized, and decisions should promote the public interest" (Bowen, 2012, pp. 64–65).

On the other hand, being rule-based implies that one follows a strict code of conduct and decision making out of a sense of duty regardless of the foreseen consequences. Kant's (1785/1964) idea of virtue is based on moral courage, the fortitude to do what is right because of a sense of duty, not because it will make us happy or help us attain some positive outcome or to avoid a negative one.

Decision makers and public relations experts should have at least a basic understanding of utilitarianism and deontological approaches to ethics, both in principles and in applications. This is not just to compare and to contrast these philosophies for their core ideas and tenets, but also to see how both might be combined and applied to certain situations. Sullivan (1989) points out that Kant (1785/1964) acknowledged the importance of consequences when applying rule-based approaches, arguing that one cannot impose a rule without looking to the final outcome of that rule's enforcement.

Faced with tough crisis-related decisions, leaders and decision makers are often confronted with the need to follow standards and procedures and at the same time to adapt to unforeseen circumstances outside the bounds of predetermined rules. More germane to the AMCM, crisis managers have to stay vigilant for any warning signs that might help them anticipate unforeseen disasters, some of which might be violations of standards or rules. For example, a manager at a factory located next to a river uncovers a slight leakage of toxic waste into the river. The seepage is slightly above industry, state, and federal standards. Even though this leakage is a violation of standards, fixing this problem would cost the factory a considerable amount of money and affect the value of the company's stock. From a deontological perspective, neglecting to fix the leakage might be an ethical violation, especially if the company was aware of the breach and chose to do nothing. But suppose the factory's leakage into the river was within acceptable standards of river contamination (i.e., not a violation of rules), but that the river contamination was reported to adversely affect the community's drinking water supply, and that reports of sicknesses linked to the contamination were starting to surface. In this case, the factory is not in violation of any industry rules, but their unwillingness to fix their leakage problem has consequences for the surrounding community. In this case, failing to fix the problem is a consequential ethical breach.

Some issues are especially conducive to crises, such as potential labor disputes, unsafe working conditions, natural disasters, and organizational misconduct. Regardless of the contributing factors, issues need to be handled with their ethical and moral implications in mind. The multiple interests and factors contributing to crises might make it difficult to predict the issues that can lead to disaster (Bowen, 2012). Nevertheless, we review common areas in which ethics can be applied, particularly from an anticipatory standpoint.

Bioethics

Bioethics concerns the advances in medicine and biology and their ethical implications. Bioethics also relates to issues of primary care, first-response and emergency care, and other branches of medicine. Experts who are qualified to weigh in on the ethics of biology and medicine come from a variety of fields, including life sciences, medicine, mental health, law, political science, and philosophy.

Some bioethical breaches involve the inappropriate or dangerous use of biological material. Such was the case in 2003, in Lubbock, Texas, when 30 vials of plague bacteria, or *Yersinia pestis*, went missing from the Texas Tech University Health Sciences Center (TTUHSC). Thomas Butler, a researcher at TTUHSC, was charged with the illegal transport of these vials into and out of the United States (KCBD.com). Butler was accused of endangering public safety, illegally smuggling bacteria, and violating federal law. Health and law enforcement officials contended that Butler's actions could have spread public panic of a bioterrorist threat across the region, but that it was curtailed quickly by local, state, and federal law enforcement. Because such cases are always a possibility, universities and agencies at all levels require that biological research facilities adhere to regulations that specify the proper use of biological material. By maintaining accountability with researchers and with other personnel who have access to such material, local officials can be the first to respond when standards are breached.

Geoethics

Geoethics involves how humans occupy and utilize the earth and its resources. Also relevant are the planetary sciences and the study of how humans respond to changes in the earth, in the atmosphere, and in natural resources.

Geoethicists concern themselves with the earth's sustainability and development, ethical and unethical ways humans interact with nature, and how populations use or exploit natural resources.

One popular area of geoethical contention is mountaintop mining, which is most common in West Virginia and in Kentucky. Sometimes referred to as "mountaintop removal," this practice encompasses mining the summit or summit side of a mountain to obtain coal. Explosives are often used to dislodge dirt in order to access the coal underneath; this process is considered less expensive than other types of mining, which involve burrowing under the earth (Yan, 2010). Mountaintop removal is the most common form of mining in the Appalachian Mountain range in the eastern United States. This practice is controversial because it strips away forestation and the life that inhabits those areas, and it has the potential to destroy the natural terrain and landscape. Add to this controversy the potential damage this practice can do to surrounding areas and communities. Environmental activists and representatives from Boone County in West Virginia have attempted to lobby Washington lawmakers to protest the mountaintop mining done by Massey and other coal-producing corporations, claiming that their practices have led to air contamination and pollution of the streams.

Business Ethics

Fair practices in customer service and trade make up the study of business ethics. This area of ethics also looks at misconduct committed against consumers and against the general public. Most business ethical disasters arise from discrimination, harassment, customer deception, illegal activities, bribery, regulatory violations, and conflicts of interest (Chandler, 2012). Such ethical breaches can damage performance, decrease productivity, disrupt operations, increase the likelihood of litigation, and damage an organization's reputation or brand. Although at least 60% of employees report witnessing ethical misconduct at work, they tend to avoid reporting it out of fear of retaliation and a lack of protection for those who report the violations (Taub, 2002).

Some of the most significant crisis events can happen when business leaders violate their ethics in order to preserve profits or to avoid tarnishing their reputations. In other cases, instead of businesses coordinating with crisis managers to anticipate or to respond to disasters (Coombs, 2012), they might engage in certain practices that interfere with this coordination, such as withholding important information from law enforcement, bribing municipal officials, or avoiding communication with the public if such efforts prevent organizational misconduct from being revealed. How business leaders respond to past crises also has ethical implications because their past responses can shape public perceptions of not only those crises but also crisis events that happen in the future (Elliot, 2012).

The practices of Johnson & Johnson during their phantom recall (Olaniran, Scholl, Williams, & Boyer, 2012) of Children's Motrin can be seen as an ethical breach with the public (see Chapter 6). The fact that Johnson & Johnson decided to avoid publicizing their recall and instead to silently buy the faulty product off the shelves, infringes on the public trust and violates the assumption that a company will be transparent about its product, especially when that product fails to perform adequately or threatens the health and safety of its consumers.

Public Service Ethics

Members of the public generally expect their elected officials and public servants to be transparent and to be honest with them, despite the popular beliefs that not all of them are (Pew Research Center, 2013; Wray, Rivers, Whitworth, Jupka, & Clements, 2006). Ethics become relevant in the ways in which public servants serve their constituents. Ethics also factor in the decisions that officials make, which ideally serve the public's interests and reflect best practices. Community members typically expect their public and local officials to avoid conflicts of interest that would undermine their objectivity, not show favoritism toward friends and family when making decisions, shun taking or offering bribes that would influence decisions, and not invest in companies or property that would be affected by one's policies (Perry, 2012).

Without even considering crises, violating the aforementioned expectations would constitute unethical behavior. But committing such violations has consequences that can go beyond tarnished reputations and loss of the public trust. Ethical breaches create conditions that make crises and disasters more likely to happen. For instance, many economics experts and laypeople attribute the financial crisis of the early twenty-first century to loose risk controls, banks knowingly selling bad investments, conflicts of interest in reporting credit ratings, and other types of unethical behavior. Also, the collapsing of buildings after an earthquake might be due to the unwillingness of designers to report or to correct faulty measurements, or choosing not to fix damages in order to keep up-to-date with building codes.

Applying Ethics to the Issues and Anticipatory Crisis Management

Issues management: An important component of ethical crisis management, particularly in the pre-crisis phase, is issues management, which is the practice of ascertaining challenging issues that affect an organization or population and curtailing their effects before they become crises (Bowen, 2012). This process should be considered an important part of the overall practice of crisis management (Heath, 1997). Issues management, among other things, includes an ethical component. Simply defining an issue that could lead to a crisis is itself an ethical act (Bowen, 2012). This is because the omission of important information by an issues manager, or the failure to provide information to the people who need it, could be seen as a moral or ethical breach, or worse, as a deliberate act of deception. Therefore, people who are charged with identifying key issues must be aware of the ethical implications involved in this task.

Ideally, issues management is conducted by one or a few key individuals who are cognizant of the issues facing a community's industrial segments. Organizations are wise to enlist an issues manager to anticipate their own red flags. In turn, a CCC should employ issues managers who can serve the interests of the industries they represent. In particular, issues managers are well poised to serve in an ethical capacity, as they are often at the forefront for identifying moral or ethical situations that can breed crisis events. In fact, issues managers are often asked to, or expected to, act as ethical consultants within their organizations (Bowen, 2008, 2012). More on the specific responsibilities of issues managers appears later in this chapter.

Bowen (2012) cautions against treating issues management and crisis management as synonymous. Rather, the two concepts are linked during the pre-crisis phase. To clarify terms, issues are potential crises. Taken further, potential crises come from the consequences of a business's decisions, matters that are left unsettled (Chase, 1984), or "a contestable question of fact, value, or policy that affects how stakeholders grant or withhold support and seek changes through public policy" (Heath, 1997, p. 44). The way in which an organization responds to, or prepares for, an issue greatly affects the community and stakeholders that it serves; therefore, issues and crisis management have deep ethical implications.

At the very core of issues management is anticipation or, more specifically, scenario building. Scenario building is the method of identifying relevant crisis issues and how they might emerge and play out for the organization or community. Scenario building requires research and lessens the impact of negative outcomes (Bowen, 2012). Research is an ongoing part of issues management and is constantly revisited. Scenario building has many ethical implications. As with effective pre-crisis management, scenario building centers on how to build trust with the public and how to maintain a positive relationship with them (Bowen, 2012). Without trust, an organization's relationship with the public can end or be permanently damaged. When issues management is done with ethics and integrity, the organization can act with consideration for their responsibility to the public and the potential consequences to them.

Buchholz, Evans, and Wagley (1994) designed a public issues management system that helps issues managers prevent crises. Bowen (2012) warns that their model does not inherently include ethical guidelines; however, it does provide a framework to which ethics could be applied. Instead, Bowen (2012) proposes an updated version of the Buchholz et al. model that takes into account Olaniran and Williams's (2001) emphasis on anticipation. Her adaptation also provides an ethical application, and it is outlined here:

1. Monitor the environment to identify public issues and trends in public expectations.
2. Conduct research and analysis using anticipatory crisis scenarios.
3. Determine relevant facts, research, and information using ethical tests.
4. Evaluate potential impacts and set priorities using ethical analyses.
5. Develop a strategy based on the ethical course of action.
6. Implement strategic communication.
7. Evaluate its effectiveness.

Bowen (2012) points out that although monitoring the environment is retained as the first step, research should take place shortly thereafter so that the extent and impact of the issue can be established, and so that information needs and issue priorities can be determined.

Bowen's (2012) system illustrates how ethics plays a role in some steps more so than others. However, the consideration of ethics is inherent throughout the process of issues management, regardless of the specific system. For instance, in the first step, one scans and monitors the environment within and outside the organization for clues that could signal trouble. For instance, an organization might scan the Internet and news publications for information about the organization and why it is getting public attention. The media can also be monitored for issues that are relevant to the organization. Bowen (2012) advocates using a utilitarian approach to monitor the environment and to scan for relevant issues. Such an approach encourages one to consider which segments of the public bear the greatest consequences of a proposed action, in both the impact and number of outcomes, which can influence how priorities are set. The publics and issues receiving higher priority do so because they stand to endure the greatest number or level of consequences. Such actions being considered might be the choice to do business with a particular organization or to make changes as dictated by governmental regulations. In general, the issues manager can determine which public segments or stakeholders are most affected by the positive and/or negative consequences of a proposed action.

Bowen's (2012) second step is conducting research and building crisis scenarios. Bowen invokes the categorical imperative when discussing the ethics of conducting research and analysis. This approach involves a decision based on rationality, which is ideally free of bias and driven only by a priori principles, rules, and morals. Bowen (2012) illustrates:

> The issues manager is required to detach him or herself from personal connections to the situation and analyze it from a 360-degree objective approach involving all viewpoints—both internal to the organization and of external publics.... The most rationally compelling moral principles that emerge from the analysis are the ethical course of action to follow. (pp. 65–66)

The application of a deontological approach implies that the organization does not privilege its viewpoint over those of others or of the public. Furthermore, an ethical guideline for fact-finding should lead to one or more universal truths rather than to the desires of management or the stakeholders. According to Kant (1785/1964), a universal truth or moral principle is one that any rational decision maker can comprehend and appreciate. A universal rule might be honesty; it is something that most people expect from others. In fact, individuals often assume that others are being honest with them until it is proven, or suspected, to be otherwise.

The third step is to determine relevant facts using predetermined ethical guidelines. Bowen (2012) argues that this step calls for Kant's categorical imperative, also referred to as Kant's "law of justice" (1785/1964). This law requires that individuals are treated not as a means to an end, but as the end themselves. To put this in a crisis context, the public should be viewed as the most important aspect of crisis management, rather than as targets to be manipulated or persuaded in order for an individual or organization to achieve a hidden mission. This idea puts the interest of the public—not the organization in crisis—at the forefront. Kant's (1785/1964) law of justice requires that issues managers look at all sides and interests of an issue, and not to prop up those of any group over others. The main concern is to avoid pursuing one's own self-interests when deciding what facts and issues are relevant to the organization. Some of the questions to ask: Does this proposed action maintain the dignity of all involved? Do members of the public have access to the information they need? What are all of the alternative points of view regarding this issue?

Fourth, in evaluating potential impacts on a community or on an organization, Bowen (2012) recommends a *rule* utilitarian approach in order to set priorities. Unlike Kant's (1785/1964) law of justice, which adheres to a preset standard or set of rules, utilitarianism generally places a focus on the end result of a decision or action, specifically one that does the greatest good for the most people involved. Bowen (2012) explains:

> In the organizational sense, a pre-crisis rule utilitarian analysis of potential consequences can provide certain classes of actions that are deemed to create benefits and others that are deemed unwise because they produce a mixed ratio of benefit to harm or of negative consequences. (pp. 69–70)

Inevitably, a utilitarian approach has the potential to harm certain people, groups, or population segments. If any harm from an action or decision is anticipated, it must be justified as promoting the overall good of the community, despite its negative effect.

Step five is to develop a strategy that ethically addresses the identified issue, which is based on act utilitarianism. This goes a step further than rule utilitarianism by requiring the decision maker to consider all of the potential consequences of the action being considered. This particular philosophy might have its limitations because it is not always possible to foresee all consequences. Nonetheless, this step is important for issues managers because forecasting the conditions that make a crisis possible is part of the job. Bowen (2012) further explains that "the potential strategies for resolving the pre-crisis issue … can be considered individually, in all of their specificity, with regard to their potential consequences" (p. 71). Each potential decision should be weighed separately to ascertain its potential outcomes. Although consequences and outcomes are sometimes difficult to predict, "a sizable amount of reliable research based on a variety of data gathering methodologies that triangulate data on the issue ... can help to offset the uncertainty prevalent in act utilitarianism" (Bowen, 2012, p. 72).

The sixth step is to implement strategic communication. Rather than being purely manipulative or designed to elicit desired outcomes in the targets, a strategic and ethical message should be based on moral goodwill. Besides achieving desired responses from the public and alleviating any crisis or threat, strategic crisis communication should reflect an obligation to do the right thing. The absence of such an intention implies that the communication should not even occur. Otherwise, acting only on the best interests of the organization would be unethical. Therefore, an organization communicating strategically and ethically must

> (1) want to actively prevent the crisis and (2) resolve any dilemmas or issues because a moral principle based on rational analysis indicates that anyone in a similar situation would conclude that [the moral] way to resolve the issue is the right thing to do. (Bowen, 2012, p. 73)

When designing and implementing strategic communication, an organization ultimately should act as a goodwill agent who is responsible to the publics it serves.

The final step in Bowen's (2012) model is to evaluate effectiveness in maintaining the public's interest. Issues and crisis managers should consider this a perception check with the public. Credibility is in the eye of the beholder—the public's eye. Evaluation research helps determine whether affected publics are satisfied with the measures taken. This helps keep the end user in mind. Also, the public needs to have an understanding of the organization's perspectives, which aids in transparency (Bowen, 2012). Research conducted by the CCC can be designed and conducted to gauge its effectiveness. Such methods can include surveys that can collect data from a large number of respondents in a short amount of time. Researchers might also conduct more in-depth interviews and focus groups at the CCC to gather more detailed information or to get clarification on survey results. The results from this research effort can enable the CCC and affected decision-makers to refine their ethical strategies and be better prepared for disasters they anticipate, and at least be better at predicting potential outcomes and consequences.

Ethical audits and analyses: Understanding the key functions of issues management enhances one's appreciation of integrity management, which is a system that is put in place to ensure that individuals will act in ways that follow legal regulations as well as uphold the organization's reputation, image, and professionalism (Chandler, 2012). Integrity management primarily involves communicating expectations of ethical behavior to employees and making sure those expectations are understood and met. Although ethical behavior implies never knowingly doing harm, it can be difficult to know what specific behaviors might violate this tacit code. For these reasons, an occasional integrity analysis is needed to check employee and personnel perceptions of predetermined expectations of ethical behavior. Chandler (2012) suggests key questions in conducting this analysis:

- On what criteria do you base confidence in your company's integrity continuity?
- Do all company personnel know how to act or to behave ethically and appropriately in all situations and contexts?
- Do employees know the rules for each situation that may arise?
- How does the company know that the employees have this information?
- How do you know, and how certain are you, that your employees are acting ethically?
- How do you define the desired ethical behavior?
- How do you implement the criteria so that they will be accepted and followed by the employees?
- Do your employees know where to turn for resources and for support for acting ethically?
- Are your employees aware of how to report ethical issues?
- Does your performance appraisal process specifically reward ethical behavior and punish unethical behavior? (p. 91)

A CCC can be a valuable source for assisting organizational leaders in creating ethical criteria, in establishing reasonable rewards and sanctions, and finding assessment tools to help answer some of these questions. Such a checklist should be also implemented with CCC personnel.

A more detailed way to address the aforementioned questions is the Ethical Conduct Audit (ECA) (Chandler, 2000), which is an evaluation that can be used to "gather information, establish reporting channels, assess culture, examine Perceived Reward Systems, be alert to warning signs, and identify patterns of potential misconduct" (p. 92). Such a system can be managed by personnel associated with the CCC who would not be connected with the organization in question, thus decreasing the chance for conflict of interest and for enhancing assessment objectivity.

The ECA provides a framework that can be adapted according to the organization's needs. In essence, it involves a SWOT analysis, or an assessment of an organization's ethical *strengths*, *weaknesses*, *opportunities*, and *threats*. The strengths and weaknesses reflect the factors that are internal to the organization, and the opportunities and threats originate from outside the organization. According to Chandler (2012), most ECAs have the following goals:

- Identify relevant ethical issues that are relevant to the organization.
- Evaluate current practices and benchmark them.
- Determine compliance levels with organizational standards, rules, and regulations.
- Review the organization's current ethics training program.
- Observe and measure new ethics initiatives (e.g., rules, policies, and rewards) for their impacts on the organization.
- Complete gaps in knowledge or decisions on changes recommended by previous ECAs.

Most ECAs require a multimethod approach, which might include surveys, observational methods, focus group sessions, or in-depth interviews. Assessors can also examine company documents, manuals, guides, and other materials from the organization. These information-gathering methods can be assigned to CCC personnel not connected to the organization but who have qualitative and/or quantitative research expertise.

One key benefit of an ECA is that it can help identify ethical vulnerabilities that are specific to that organization, such as conflicts of interest or unintended incentives for unethical behavior. Additionally, it can allow for the identification of ethics-related warning signs before they lead to actual disasters: "One unique aspect of the ECA is that it … may help provide an early warning for potential unintended consequences of the various decision processes" (Chandler, 2012, p. 93).

Models of analysis that facilitate a quick response when an ethical crisis does occur are ideal. One such model is RADAR (Brewer, Chandler, & Ferrell, 2006). The RADAR (recognize, avoid, discover, answer, recover) model is intended to help decision makers recognize and mitigate the risks of ethical catastrophes. The first step, *recognize*, involves assessing risk, forecasting relevant threats, and anticipating potential lapses in ethical standards and how to overcome them. The second step, *avoid*, requires one to develop one or more leadership strategies that are consistent with the ethics policies and codes of conduct already in place. Those in leadership positions can help facilitate needed training in policies, help construct compliance programs, and establish objectives for training sessions. Moreover, it is often up to the leaders to make sure those objectives and codes are communicated clearly to personnel. To *discover* means to employ those previous steps to ensure that warning signs of ethical disasters are exposed early before too much damage can occur. Discovery involves scanning for problems, being open to feedback about potential ethics issues, checking behavior, and engaging in active communication with cohorts, subordinates, and those at the top. The fourth step is *answer*: "When ethical breakdowns occur it is critical that management quickly responds to the uncovered situation in order to prevent an issue from growing into a scandal or to keep a scandal from becoming an ethical misconduct disaster" (Chandler, 2012, p. 94). For instance, an answer to an ethical concern can be a response protocol for a conflict-of-interest violation. Finally, the *recover* step depends a great deal on the organization's preparedness for potential ethical disasters in the first place. Although some ethical scandals can be extremely overwhelming and consuming, a well-prepared organization is able to bounce back, repair, and restore its reputation. Whatever the response, it needs to be the kind that works toward resuming productivity, restoring shareholder value, and reestablishing customer or public confidence and trust.

Ethical Standards and Policies for the CCC

This chapter has been devoted to discussing ethical perspectives and to discussing practical and theoretical approaches to applying ethics to avoid or to respond to disaster. The remainder of this chapter was devoted to specific guidelines for setting up ethical standards within the CCC and strategies that a CCC can employ when assisting one or more community entities in handling virtually any local crisis. What follows is a suggested ethical code of conduct that could be adopted by a CCC and can serve as a general outline of an ethics division that could be housed in, and operated by, a CCC.

The CCC Ethical Code of Conduct

An ethical code of conduct can enable a CCC to communicate clear guidelines for internal staff and community members that collaborate with the CCC on behalf of their organization (e.g., liaison from a local business, representative from the fire department, or physician-representative from a local hospital). Table 8.1 summarizes a typical ethical code of conduct.

Table 8.1: CCC Ethical Code of Conduct

1.	Adhere to crisis and risk management industry standards
2.	Uphold the mission, vision, and purpose(s) of the CCC
3.	Maintain the highest levels of skill and knowledge required by the CCC
4.	Avoid or report potential conflicts of interest
5.	Avoid and report discrimination based on sex, race, nationality, religion, sexual orientation, and other cultural classifications
6.	Maintain full disclosure with the public about the CCC's purpose and practices
7.	Maintain individual and organizational confidentiality whenever appropriate or legally required
8.	Treat all personnel, community members, and organizations with dignity and respect
9.	Communicate clearly the rights and responsibilities of clients

Adhere to crisis and risk management industry standards. Industry standards reflect specific practices of crisis management, crisis communication, and risk management, and how professionals are trained and certified. Such standards can include those practices and policies outlined earlier in this chapter, such as clear communication of guidelines, early communication about, and response to, disaster as it occurs, ethical relations with local media, and honest communication with the public.

Uphold the mission, vision, and purpose(s) of the CCC. Following this code depends a great deal on how clear the mission, vision, and CCC purpose is to its personnel and to the community being served. The mission statement should indicate who the CCC is as an entity, why it exists, and what functions it serves. The vision articulates the short- and long-term goals of the CCC and the general direction in which it is headed in relation to the community it serves.

Maintain the highest levels of skill and knowledge required by the CCC. This guideline pertains to skill sets specific to crisis communication, issues management, and risk communication about issues unique to the community.

For example, CCC personnel should be aware of actions required in response to certain types of crises (e.g., natural disasters vs. terrorist threats). Knowledge and skills also include communication-based practices, such as when informing the public of practices that have been established by crisis experts.

Avoid or report potential conflicts of interest, which means making sure those efforts to maintain or to reestablish security and safety are not undermined by the potential of some parties to gain from the crisis, or the response to that crisis. For example, if a local manufacturing company is distributing a product that is causing harm and needs to be recalled, any shareholder or employee of that company should not be on the team that informs the public. There is too much of a possibility that, in order to restore or to protect the company's image, such persons might not be as honest with the media or with the public as those who have no personal or occupational stake in the company.

Avoid and report discrimination. Upholding equity in treatment and opportunity should apply not only to CCC personnel but also to representatives of partnering agencies and the community members being served. Nondiscrimination clauses are not all created equal. While the usual statements ban discrimination based on race, sex, nationality, or creed, some do not openly include protections based on other classifications such as sexual orientation. The CCC should not constrain itself just to statements mandated by the state or federal government, but nondiscrimination clauses should be written to ensure protection for all individuals who potentially suffer discrimination.

Maintain full disclosure with the public about the CCC's purpose and practices. The functions of the CCC, the people and organizations they serve, the entities with which they partner, and past decisions and actions, should never be hidden from public view. Certain CCC personnel should be in charge of maintaining records that can be made available publicly through written and electronic form (e.g., websites). Because it often works in tandem with government-level agencies, the CCC should consider itself accountable to the public, who have a right to their information. As a show of good faith, the CCC can maintain regular blogs or news stories about its goings-on, both during a crises and times of stability.

Maintain individual and organizational confidentiality whenever appropriate or legally required. Maintaining confidentiality or anonymity should not be conflated with hiding unethical or malevolent practices. On the contrary, all individuals and organizations should be transparent about their actions and be held accountable by the public. However, there are instances when individuals' or agencies' identities should be kept private, such as when there are victims of attacks or natural disasters, especially if it is at the victims' request. Another example might be during an investigation of a terrorist threat, when revealing a suspect's identity might tip off perpetrators and thus compromise the investigation. The key to *appropriate* and/or *legal* confidentiality requires working closely with community agencies that can inform the CCC about legal, state, and federal requirements regarding identity disclosure.

Treat all personnel, community members, and organizations with dignity and respect. Though not typically required from a legal standpoint, a statement affirming the dignified and respectful treatment of all persons involved is a show of goodwill toward the community and toward the CCC personnel. Such a statement might be placed alongside a mission/vision statement and should be easily visible on promotional information disseminated to the public. Because it has already been stated that rewarding appropriate behavior can be praiseworthy, the CCC might consider ways in which to recognize and reward individuals and organizations that promote dignity and respect.

Rights and responsibilities of clients served by the CCC. Individuals and groups who receive assistance from the CCC deserve to know the rights to which they are entitled. They also deserve to be informed of their responsibilities, such as information they need to provide or tasks they need to perform. Providing these dual sets of statements illustrates how the CCC and their clients have responsibilities toward each other, thus having a mutual influence on each other.

The CCC Ethics Division

In order to ensure a consistent code of ethics and reliable expectations for following them, the CCC should consider establishing its own ethics office or division. Depending on the size and reach of the CCC, and the community it serves, such a division might be composed of individuals who are solely dedicated to ethics and occupied by staff who perform other operations within the CCC. Another possibility is to involve community representatives who act as ethics board members or as representatives. If a distinct ethics division—with a group of individuals who are commissioned for this work—is not feasible, then the CCC needs to provide a physical home base for a community crisis ethics committee. Such a committee can have as many people as is appropriate for the size of the community and include representatives from various offices or agencies (e.g., law enforcement, medical industry, business, and education).

The ethics committee or division can serve a variety of functions, one of which is to receive ethics complaints or reports of ethics violations. If the violations or complaints are deemed to have merit, then the division can refer the problem to an agency that can more directly deal with the breach. For example, if the CCC receives a complaint about a local business's failure to report a dangerous chemical exposure, the ethics division can determine which community resource might be consulted to address the problem (e.g., local environmental agency, law enforcement, or the Better Business Bureau office). Such an action might curtail any catastrophic event that might emerge from the breach.

The ethics division is only as good as the training and knowledge of its members to handle ethics issues. The CCC can partner with local educational institutions to provide ethics and moral philosophy training, which is worth the investment to prevent crises or moral lapses (Bowen, 2012). In addition to training that might already exist at the college or university level, the CCC can provide resources or help facilitate ethics training in the local systems of higher education.

Another possible function of the ethics division is to help organizations and agencies identity and train issues managers within their own ranks. These experts are specific to the fields and industries prevalent within the community. The issues manager not only adjudicates how to handle a crisis but is also able to research the facts surrounding actual and potential ethical crises (Bowen, 2012). Issues managers' jobs are ethical in nature because they carry a significant responsibility in deciding what information is pertinent, who needs access to that information, and how the research to gather that information will be conducted. As Bowen (2012) points out,

> [Issues management] requires a high level of discernment because if key facts are omitted, an ethical violation of the public's right to relevant information has already been committed—even before any decisions are made on how to handle the issue. Further, those publics might perceive a misjudgment by the issues manager as an intentional concealment on the behalf of the organization, or worse, as an intentional deception. (p. 61)

Because of the ethical implications, the CCC's ethics division should help organizations formulate their issues management strategies.

As with the CCC in its entirety, the ethics division can provide input for the management or recovery of an organization's reputation within the community, their customers, and their stakeholders. As this chapter has previously discussed, many organizational crises result from ethical lapses or unethical decisions. When this happens, the offending organization needs to regain the trust of the public, which is otherwise a fundamental building block of any organization (Dietz & Gillespie, 2012), because an organization's success or profitability depends a great deal on the trust of the public. Consulting the ethics division of the CCC can be a sound first step in rebuilding that trust.

Finally, the ethics division should not only be tasked with serving as the watchdog of the community; it also needs to conduct checks on the CCC itself. For instance, the division can enlist people outside the CCC to conduct regular audits of communication, press releases, newsletters and other outgoing messages. Such checks should determine accuracy, honesty, and potential impact on the receivers.

Conclusion

This chapter covered the ethical issues involved in crisis communication from the perspective of CCC operations. The section on applied ethics reviewed predominant ethical perspectives and how they relate to crisis management. More specifically, bioethics, geoethics, and business and public services ethics were discussed. The second section, which is linked to the AMCM, addressed the importance of issues management and provided sample systems of ethical analysis and audits. Finally, the chapter made specific recommendations for a CCC code of conduct and a setup for an ethics division contained with a CCC. This chapter's main purpose was to provide a rationale for maintaining ethical integrity and to describe how a CCC can aid in the maintenance of ethics standards of the community entities with which it partners. Ethics frameworks that reward and encourage ethical behavior, provide simple guidelines, and coincide with individuals' ethical frameworks tend to be the most enduring (Chandler, 2012).

Go to the following resources for more information on ethical decision-making:

- Proposed rural healthcare ethics agenda: http://jme.bmj.com/content/33/3/136.short

- Organizational ethics decision-making toolkit: https://www.ache.org/ABT_ACHE/EthicsToolkit/Ethical_Decision_Process.pdf

References

Bowen, S. A. (2008). A state of neglect: Public relations as "corporate conscience" or ethics counsel. *Journal of Public Relations Research*, *20*(3), 271–96.

Bowen, S. A. (2012). The ethical challenges of pre-crisis communication. In B. A. Olaniran, D. E. Williams, & W. T. Coombs (Eds.), *Pre-crisis planning, communication, and management: Preparing for the inevitable* (pp. 57–78). New York: Peter Lang.

Brewer, L., Chandler, R. C., & Ferrell, O. C. (2006). *Managing risks for corporate integrity: How to survive an ethical misconduct disaster*. Mason, OH: Cengage Learning.

Buchholz, R. A., Evans, W. D., & Wagley, R. A. (1994). *Management responses to public issues: Concepts and cases in strategy formulation* (3rd ed.). Upper Saddle River, NJ: Prentice-Hall.

Chandler, R. E. (2000). Managing ethical and regulatory compliance contingencies: Planning and training guidelines. *Contingency planning and management 2000 proceedings*. Flemington, NJ: Witter.

Chandler, R. E. (2005). Avoiding ethical misconduct disasters: Strategic planning for organizational "integrity" continuity is essential for avoiding, mitigating, and surviving organizational scandals and (un)ethical disasters. *Graziodio Business Review*, 8(3). Accessed October 26, 1012, http://gbr.pepperdine.edu/2010/08/avoiding-ethical-misconduct-disasters/

Chandler, R. E. (2012). Recognizing, anticipating, and preventing ethical misconduct disasters. In B. A. Olaniran, D. E. Williams, & W. T. Coombs (Eds.), *Pre-crisis planning, communication, and management: Preparing for the inevitable* (pp. 79–97). New York: Peter Lang.

Chase, W. H. (1984). *Issue management: Origins of the future*. Stamford, CT: Issue Action Publications.

Coombs, W. T. (2012). Crisis communication and its allied fields. In W. T. Coombs & S. J. Holladay (Eds.), *The handbook of crisis communication* (pp. 54–64). West Sussex, UK: Wiley-Blackwell.

Dietz, G., & Gillespie, N. (2012). The recovery of trust: Case studies or organizational failures and trust repair. Institute of Business Ethics. http://www.ibe.org.uk/userfiles/op_trustcase studies.pdf

Elliot, J. D. (2012). How do past crisis events affect publics' perceptions of current events? An experiment testing corporate reputation during an adverse event. In W. T. Coombs & S. J. Holladay (Eds.), *The handbook of crisis communication* (pp. 205–20). West Sussex, UK: Wiley-Blackwell.

15 count indictment against Thomas Butler. (2003, April 10). KCBD. Accessed October 30, 2012, http://www.kcbd.com/story/1227002/15-count-indictment-against-thomas-butler?-clienttype=printable

Heath, R. L. (1997). *Strategic issues management: Organizations and public policy challenges*. Thousand Oaks, CA: Sage.

Kant, I. (1785/1964). *Groundwork of the metaphysics of morals*. H. J. Paton (Trans.). New York: Harper & Row.

1984: Hundreds die in Bhopal chemical accident. (2005, December 3). BBC Home. Accessed October 29, 2012, http://news.bbc.co.uk/onthisday/hi/dates/stories/december/3/newsid_2698000/2698709.stm.

Olaniran, B. A., Scholl, J. C., Williams, D. E., & Boyer, L. (2012). Johnson and Johnson phantom recall: A fall from grace or a re-visit of the ghost of the past. *Public Relations Review, 38*, 153–55, doi:10.1016/j.pubrev.2011.08.001.

Olaniran, B. A., & Williams, D. E. (2001). Anticipatory model of crisis management: A vigilant response to technological crises. In R. L. Heath (Ed.), *Handbook of public relations* (pp. 487–500). Thousand Oaks, CA: Sage.

Perry, D. L. (2012). Ethics in public service. Markkula Center for Applied Ethics. Santa Clara University. Accessed October 30, 2012, http://www.scu.edu/ethics/publications/submitted/Perry/service.html

Pew Research Center (2013). How Americans view government: Deconstructing distrust. Pew Research Center for the People and the Press. Retrieved August 3, 2013, from http://www.people-press.org/1998/03/10/how-americans-view-government/

Sinnott-Armstrong, W. (2012) Consequentialism. *The Stanford encyclopedia of philosophy*, E. N. Zalta (Ed.). Accessed October 30, 2012, http://plato.stanford.edu/archives/win2012/entries/consequentialism/

Sullivan, R. J. (1989). *Immanuel Kant's moral theory*. Cambridge: Cambridge University.

Taub, S. (2002). Crisis of ethics: Ethics officers predict a new wave of corporate scandals. *CFO.com*, Accessed October 26, 2012, http://www.cfo.com/article.cfm/3005220?f=search.

Wray, R., Rivers, J., Whitworth, A., Jupka, K., & Clements, B. (2006). Public perceptions about trust in emergency risk communication: Qualitative research findings. *International Journal of Mass Emergencies and Disasters, 24*(1), 45–75.

Yan, S. (2010, March 12). In West Virginia, a battle over mountaintop mining. *Time*. Accessed October 30, 2012, http://www.time.com/time/health/article/0,8599,1971709,00.html

Case Study

Private Police Forces and Government Ethics

Source: Robert Wechsler. http://cityethics.org/node/4447

What are the government ethics implications of private security when it goes beyond protecting specific businesses, malls, universities, and gated communities, becomes an adjunct to, or replacement of, an ordinary police force, and is done in conjunction with the public police force and, often, using off-duty public police officers?

Favoritism: One problem is that such private forces generally protect the wealthiest neighborhoods. Setting up a neighborhood force with the support of a city or county government is little different from offering the government a gift of funds for the police force, but insisting that they all be spent in one wealthy neighborhood. This sort of favoritism sends a message that one's police protection depends on one's wealth, a form of favoritism that is not an ethics violation, but which undermines trust in government equivalent to the provision of worse schools in poor neighborhoods, without the alternative of busing, magnet schools, vouchers, or other means of equalizing education. Privatizing an entire police force is less problematic, at least if the officers are subject to government ethics and other laws.

Conflicts of interest: Another problem arises when off-duty public police officers work for private police forces in these wealthy neighborhoods. Although their work is similar, these officers are wearing two hats and working for two bosses, which may have different interests and goals. In fact, these different interests and goals are sometimes explicitly stated in state and local laws, rules, and regulations, for example, in Illinois's Private College Campus Police Act of 1992, which states that it is intended to protect not only the "students, employees, visitors and their property," but also the "interests of the college or university, in the county where the college or university is located." There are going to be a variety of situations where off-duty officers working for a university are conflicted, and it won't be easy to withdraw from participation or to otherwise cure the conflict.

A different sort of conflict of interest arises when a city with a private police force is considering limits on outside work by police. It is difficult for a city council to pass a law that would limit a private police force's ability to hire off-duty officers, when council members know that the force protects most of their major donors.

Transparency: A third problem involves secrecy: private police forces are usually not required to have the same level of transparency as public police forces. This could be required, but it does not appear to be common. Especially since constitutional safeguards do not generally apply to private police, it is especially important that there be full transparency.

A New Orleans example: A recent *New York Times Magazine* article looked at an especially problematic private police force for the wealthy French Quarter of New Orleans. According to the article, the force was started by one individual, who happened to have made a great deal of money from city waste removal contracts after Hurricane Katrina and who appeared to be loaded with conflicts of interest.

One conflict involves the fact that, with city government support and the use of off-duty public police officers, he is using his short-term investment in his own neighborhood (after his house was burglarized) to test an app that he appears to want to sell to neighborhoods across the country.

Another conflict involves the head of the state police, who, according to the article, once visited the individual at a resort he owned in the Bahamas. A high-level official such as this should not be involved in any way with a friend's public-private venture. And, yet, he is publicly supporting the individual and invited him to attend a state police sting operation.

Another issue involves ads that the individual paid for, attacking the mayor for allowing a crime wave. The attack ads helped push the mayor to allow the individual to get city support for his neighborhood's experiment, even though the growth of crime occurred throughout the city.

Yet another issue involves gift certificates the individual gave to particular officers he felt did a good job. Not only is this similar to an illegal gratuity to a police officer, but it is part of a personalization that can accompany such privatization. The individual became involved not only in managing the force, but in setting city policy, for example, by pushing for a joint sweep of panhandlers in the French Quarter. Also, the individual's "hunger for credit, his disregard of regulations, his habit of leaking sensitive information that could make it harder to prosecute suspects—all this had resulted in more than a few heated arguments" with the city government and its police force.

These are just a few of the government ethics issues that may come from establishing private police forces in wealthy city neighborhoods.

Discussion Questions:

1. How is favoritism a violation of ethical principles, according to the ethical approaches discussed in this chapter?
2. How should the CCC partner with public and private police forces? Should there be specific rules that apply to private police or security officers?

· 9 ·

PUTTING THE CCC PLAN INTO ACTION

Chapter Highlights

- The mission statement: What an organization does, why it exists, and how it contributes to the greater community.
- The vision statement and short- and long-term goals
- The general operations of the CCC
- The crisis communication plan

This chapter provides a blueprint that local CCCs can use to create their own crisis communication plans. This framework assists the CCC and its partners in utilizing the appropriate community sectors during any time of crisis. This chapter largely focuses on what the CCC can do to encourage, educate about, and facilitate effective communication regarding crisis management policies and procedures, but the main objective is to introduce tools, protocols, and templates that the CCC can provide for agencies and organizations as they develop their own individualized crisis plans. Furthermore, the notion of anticipation is reviewed and applied to common crisis situations that many CCCs might face.

This chapter is organized into three sections. The first section provides a rationale and guidelines for formulating the mission and vision statements for the CCC. The second section outlines the components of a CCC's plan of operations. This section does not advocate for one plan executed across all contexts. Rather, each CCC should be able to add to the plan and to adjust it to fit the needs of their particular community and adjust it according to the nature of the crises that pose the greatest threat. The third section deals specifically with the communication-based planning and procedural assistance that the CCC can provide for community agencies and organizational partners. The chapter also outlines a risk and crisis management planning process that organizations can implement, with specific focus paid to communication-related issues, such as media and public relations, effective interorganizational networking, and education and training.

Mission and Vision Statements

A CCC's mission and vision have internal and external functions. Internally, these statements reinforce the CCC's purpose and goals, thus helping to maintain consistency in its operations. They also provide a framework around which personnel can better understand the goals of the CCC and how the departments and personnel are related to each other. Externally, the mission and vision communicate the CCC's purpose and enhance its credibility. Such statements should be visually apparent as clients and community members enter the CCC's doors, and they should be clearly visible on all printed and electronic material, and on the local CCC's website.

Mission Statement

A mission statement communicates to the public what an organization does, why it exists, and how it intends to contribute to the greater community (Abrams, 2003). More specifically, the mission statement serves to guide the CCC's actions, to present its general goals, and to provide a template for decision making (Hill & Jones, 2008). A noncommercial mission statement should contain the following elements:

- The CCC's aim
- The CCC's main stakeholders

- How the CCC provides value to its stakeholders
- Why the CCC exists
- How the CCC differs from other community and municipal agencies

Given these components, here is a sample mission statement:

> The Crisis Communication Center of Pleasantville seeks to be the primary source of communication and media support for all organizations and agencies in the community before, during, and after crisis events and disasters. In serving as this community link, the Crisis Communication Center provides training for anticipating and managing disasters, as well as a venue and framework to assist organizations as they manage crises that do occur.

This statement would not necessarily apply to all types of CCCs, but can serve as an example for how a mission statement articulates its reason for existing, how it differs from other organizations, and which stakeholders it serves.

Vision Statement

While the mission statement provides a general picture of the organization's goal, the vision statement projects into the future with short- and long-term goals (Lipton, 1996). Whereas the mission describes what the organization does, the vision illustrates what the organization wants to be and how it will achieve its mission. Moreover, a well-written vision statement motivates the CCC's daily operations and informs its strategic plan. For example, based on its mission, the Pleasantville CCC describes its vision: "The Pleasantville Crisis Communication Center strives to be the primary source of communication tools and resources when tragedy strikes." This vision is aspirational in nature in that it lays out the most important goal of the company. However, Hom (2013) warns that the vision should not be confused with a business plan, which describes how the company will achieve its primary goals. Instead, the vision illustrates the main focus of the company, which will enable the development of more specific strategies.

Hom (2013) provides the follow guidelines for writing an effective vision statement:

- Write the vision with enough vagueness so that it can be adapted over time, just like most goals are adapted to fit the changing needs of the organization.

- Use the present tense.
- When describing the CCC's goal(s), project five to ten years in the future.

In addition, Hom (2013) and Kirkpatrick, Wofford, and Baum (2002) suggest using imagery that inspires emotion and passion. Imagery has been linked with charisma and with leadership (Kirkpatrick et al.). In general, the CCC's mission and vision do not just provide an image that is conveyed to community members and stakeholders. These statements can be used by key personnel to articulate the CCC's strategic plan to set long- and short-term goals, promote positive change, help recruit talent, and motivate members of the organization (Lipton, 1996).

General Operations

Personnel

CCCs will vary in the type and number of people that make up its operations. The size of the community and the availability of resources and funds will be significant factors in how many individuals comprise full-time CCC staff. CCC personnel might also include unpaid individuals who occupy sectors within the community. Therefore, this section recommends personnel that exist within and outside the CCC.

Board of directors: Informing the internal personnel should be a board of directors who are appointed or elected to oversee all CCC operations. At the CCC's inception, the board can be appointed by the sector liaisons (described in the next section), and once the board is in place, its members appoint new members to fill vacant positions. Another option is to have members elected from the community at large through a municipal vote. Duties of the board might include, but are not limited to, determining the board's procedures and policies, selecting or electing the executive director, overseeing the availability and expenditure of the CCC's finances, and determining the salaries of the executive director and the permanent support staff (Foundation Center, 2013). The board of directors always act on the behalf of the CCC and provide guidance and suggestions to the executive director and any appointed sector liaisons (see the following section).

Executive director: The CCC should have an executive director (ED) and a support staff, which is needed for program operations. The executive director often has a role that is similar to that of a company CEO. Although the CCC is considered a nonprofit organization, its director might also take on the title of CEO or president (Fram & Brown, 1995). The ED's typical role is to oversee and to participate in the development of the CCC's strategic plan. In addition, the ED is responsible for the day-to-day operations and executive leadership skills that help the CCC determine its mission and vision. Despite the amount of authority the ED has, he or she is accountable to the board of directors, who gives him or her the authority to run the CCC. In accordance with the bylaws, the ED may report to the board on an annual, semi-annual, or quarterly basis. In coordination with the board, the ED needs to be informed of all issues pertaining to personnel, budget, assets, internal and external resources, and other issues that affect operations. Although the ED answers to the board's authority, the ED can decide whether or not to execute the board's ideas and suggestions. The ED plays an important leadership role within and without the CCC, mentoring staff and liaisons, motivating team members, and providing direction during crises and when the CCC is developing its strategic plan (Hill & Jones, 2008).

Support staff: The support staff can be full-time, part-time, or volunteer, depending on the funds available and the need determined by the ED and by the board of directors. Support staff answer phones, provide clerical support, conduct correspondence, and serve as the gatekeepers and first faces for visitors to the CCC. Staff should also include a webmaster or a web team in charge of establishing and maintaining the online presence of the CCC, as well as monitoring and overseeing the digital asset management system (see the next section).

Communities will vary in what they are able and willing to pay a CCC's internal personnel. Because an ED is crucial to the efficiency and effectiveness of CCC operations, it is ideal that the budget include a salary for this position. Communities or townships that do not have the funds to hire a full-time ED might be able to enlist a volunteer; however, the lack of salary—even one that is part-time—would make it difficult to ensure that the CCC has consistent, reliable, and well-qualified leadership. An ED salary can be stipulated in the bylaws. If funds are not able to pay support staff on a full- or part-time basis, these positions can be available as volunteer positions or as student internships.

CCC *liaison*: Another possible paid position would include the CCC liaison, a person within the CCC who serves as the link to the participating organizations and agencies who receive assistance from the CCC. The liaison can also act on behalf of the CCC to help partnering organizations establish their own crisis management teams, and can also advise the various crisis teams by recommending a CEO, a public relations representative, a vice president, and a manager from each division who might be called upon to address a crisis. It is important to remember that neither the CCC nor its liaison make the crisis-related decisions for the community organization; that is the organization's task. The CCC liaison or advisor also should ensure that each organization has a communication strategy that allows their crisis team members to communicate with each other quickly and efficiently when a crisis is possible or inevitable. In terms of budget, a full- or part-time salary would be ideal for the CCC liaison. If this is not possible, another option is to appoint or to elect a liaison team to spread the workload. Such a policy can be stipulated in the bylaws.

Sector liaisons: In addition to the internal staff, the CCC liaison person or team will need to stay linked with the active agencies and sectors. The CCC liaison should designate a sector liaison from each of the community's sectors (e.g., law enforcement, medical, school district, municipal office, or agricultural), whom the CCC can contact during times of crises and stability. These individuals—who are considered external personnel—should maintain regular contact with the CCC liaison and ensure that their contact information is updated. This updated information is also necessary to alert the sector liaisons to regularly scheduled training (e.g., message strategies, dealing with the media, communicating with the public, and communicating internally) that empower them to perform their roles effectively. The CCC should provide a support system that can facilitate the sector liaison's selection and training. An alternative liaison for each sector should be designated.

Spokespersons: Each sector (e.g., police, medical, and education) will likely contain one, two, or more organizations or agencies. For example, a community's law enforcement sector would consist of the local police department and the county sheriff's office. Also, the medical sector might contain two major hospital systems, individual clinics, and doctor offices. As for business or commerce, there could be several chains or small businesses that make up the commercial sector. Each of these businesses, agencies, and offices would be advised to designate a spokesperson, who represents them and who speaks on their behalf during crisis threats or actual crises. As with any good crisis management strategy, such a spokesperson is already a prominent leader in the organization, knows how to communicate with the media, represents the company to the public, and is the central person to whom questions and inquiries go during crisis. The CCC should provide a support system that can facilitate the spokesperson's selection, training, and support during crisis. An alternative spokesperson should also be designated.

These spokespersons should regularly get in touch with the CCC liaison and with the CCC in general to ensure their contact information is updated. This updated information is also necessary to alert the spokesperson to regularly scheduled training (e.g., message strategies, dealing with the media, communicating with the public, and communicating internally). All external personnel should register with the CCC by providing contact information, organizational affiliation, professional experience, and educational background. As part of registration, every external member should complete a comprehensive profile that solicits operational, objective, and personal characteristics (Mathes & Stevenson, 2011). Such a profile is shown in Figure 9.1 .

Operational characteristics	Role within his/her organization and consequent value systems Daily concerns and attitudes Knowledge of his/her technical responsibilities and assignment What he/she will need from the CCC What staff and other persons are connected through his/her duties How his/her role is affected by connection to the CCC
Objective characteristics	Education (e.g., levels, fields, years, and schools) Past professional experiences and roles Knowledge of the CCC
Personal characteristics	Self-reports of personal characteristics that could influence his/her reactions and performance Age, attitudes, and pet concerns

Figure 9.1: Comprehensive Profile for Sector Liaisons and Spokespersons

The executive director, board of directors, operations staff, and other key individuals in the community need to have a serious conversation on how the CCC should operate and how they make the CCC relevant to the community. A clear, centralized communication network of personnel appears in Figure 9.2. At the center is the CCC and its staff; they have a direct relationship with the CCC liaison, who works in-house and externally. The liaison should maintain a good working relationship with the sector liaisons, who represent the interests of their sections, and communicate regularly with spokespersons of partnering businesses, agencies, and organizations.

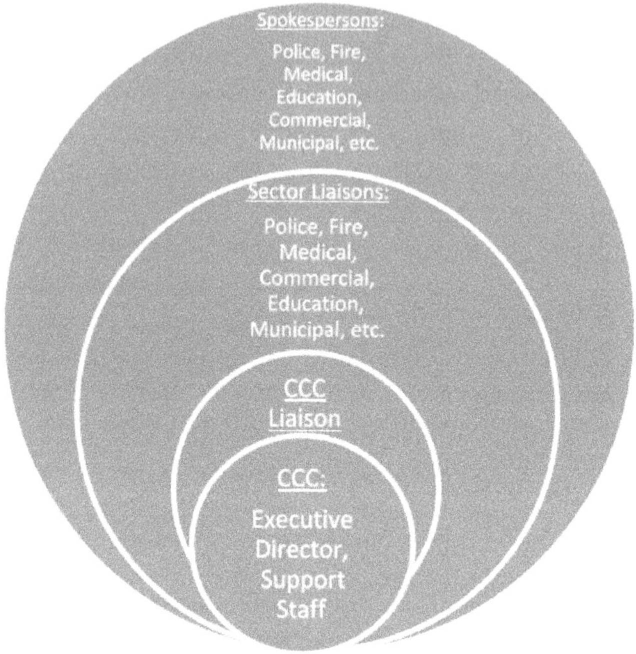

Figure 9.2: CCC Network of Personnel.

Bylaws

Bylaws are an essential component of the decision-making process for the board and for the CCC, and they lay the groundwork for governance. Although they vary, most bylaw documents contain these elements (Stephens, 2000):

- Names and descriptions of all personnel
- Meetings: location, frequency, required attendees, and requirements for meetings to be held (e.g., quorum)
- Voting procedures: election or appointing of members, on actions, who is eligible to vote, and role of proxies in voting
- Board of directors: duties, number, membership tenure, and election/appointment and removal
- Officers: roles, election and removal, and executive committees
- Records: types, inspection, writing, retention, and disposition
- Funds: raising, spending, and management
- Amendment of bylaws

The board of directors should develop and maintain their own set of bylaws, but the board should have input on the CCC's bylaws. Many online resources, for free or for a fee, provide a template with which to build a set of bylaws. Be sure to compare these services, and be knowledgeable that regulations regarding nonprofit bylaws vary by state.

Strategic Plan

A strategic plan allows an organization to define its approach and future direction, as well as to allocate its resources and to pursue an overall strategy for growth and development. To determine an organization's future direction, Armstrong (1986) suggests asking three questions: "What do we do?" "For whom do we do it?" and "How do we excel?" The strategic plan should include the organization's mission and vision statements, which set the tone for how the strategic plan helps the CCC achieve its short- and long-term goals, and how those goals will be accomplished (Wilson, 2001). Strategic plans strive for the success of an organization. While for businesses it tends to be geared toward the bottom line, a CCC measurement of success can lie in its measurable contributions to the safety of the community and being able to reach and impact diverse publics (Wilson, 2001).

Another component is the strategy or plan, which outlines the means with which the CCC will reach its goals. Planners should think of this as a roadmap, a path that moves members through the process. One aspect of the strategic plan is the timeline, which includes milestones for each task to be completed. Suppose a CCC develops a strategic plan for its first five years. The CCC can set milestones, such as hiring staff or recruiting volunteers, and enlisting a CCC liaison or team, sector liaisons, and company spokespersons. The time it takes to enlist these key individuals will vary on such factors as the size of the community, the businesses and sectors represented, and the willingness and ability of businesses to designated spokespersons. Once enough personnel have been incorporated into the plan, another important milestone is the crisis management plan, risk assessments, a communication plan, and ongoing efforts toward training and education. After initial components of the organization have been established, the CCC will need to revise its strategic plan continually to adapt to environmental changes, to keep up with new media and technologies, to enable its growth and impact, and to plan for leadership succession.

It is best if the executive director and at least most of the board of directors are in place as the strategic plan is being developed. Nonetheless, the choices made during strategic planning at the communication and management levels will be essential to the organization's success. To this end, Heath (2001) argues that public relations specialists could take part in the strategic planning as long as their role focuses more on public image and translating the identity to outsiders. The public relations role becomes less helpful if it emphasizes tactical or micro-level planning (e.g., hiring of support staff and rules regarding a quorum at board of directors meetings).

Crisis Management Plan

The crisis management plan is presented as a core component of the CCC's strategic plan. The plan is a proposed framework for agencies and organizations and is based on the Anticipatory Crisis Management Model (Olaniran & Williams, 2001). A variety of crisis situations exist, but some are more unique to specific communities. Communities should be mindful of Coombs's (2014) typology of crisis events and determine which ones most affect them. The crisis events include

- Natural disasters: Resulting from weather or "acts of God."
- Malevolence: Someone on the outside employing extreme tactics to express anger toward an organization.
- Technical breakdowns: When an organization's technology breaks down or fails.
- Human breakdowns: Disorder caused by human error.
- Challenges: Resulting from disgruntled stakeholders confronting the organization.
- Mega damage: Accidents that cause significant damage to the environment.
- Organizational misdeeds: When managers or employees behave in ways that cause harm to stakeholders, the public, or the environment.
- Workplace violence: When employees or employers commit violent acts against others onsite.
- Rumors: Spreading false information about an organization that compromises its reputation.

Not only do these specific crises have the potential to cause serious physical, emotional, and economic damage to the community, but the effects can be made worse by incompetent handling by decision makers and by spokespersons' poor relations with the media. Another sign of poor handling is not having assessed which types of crises affect a community or sector the most. Not all communities are vulnerable to the same types of crises. Understanding these implications is essential to an anticipatory approach in a crisis communication plan. Organizations that have a plan that embraces an anticipatory approach will experience enhanced crisis management capabilities. Furthermore, the CCC can emphasize to its partners that crisis management and anticipation are key, and that such plans must be institutionalized in order for the organization to respond appropriately (Sapriel, 2003), as well as to enable the CCC to offer needed support. What follows are elements of a CCC's crisis management plan. In turn, the CCC can use this framework to guide organizations and agencies to adopt their own crisis management plans.

Risk Assessment

According to the anticipatory model of crisis management, crisis managers are more successful when they proactively foresee the likelihood of a crisis. Anticipation involves scanning for all the environmental factors, technological risks, and financial vulnerabilities; these activities comprise the risk assessment (Scholl, Heuman, Olaniran, & Williams, 2007). According to the Federal Emergency Management Agency (FEMA, 2012), risk assessment is an environmental scan of the community to identify potential hazards and to determine what would happen if those hazards occurred. Methods of risk assessment vary, depending on the community needing the assessment and who is conducting it. FEMA offers a risk assessment plan through ready.gov, which involves three components—hazard identification, vulnerability assessment, and impact analysis. Hazard identification pinpoints all of the potential harms that likely put a community at risk, such as terrorism, utility outages, weather patterns, or pandemics. The likelihood and magnitude of the identified hazards will then inform the vulnerability assessment, which is a survey of the community assets at risk. Such assets include people, property and buildings, contractual obligations, businesses and industries, or current infrastructure. The extent to which the assets are vulnerable informs the impact analysis. Impacts are outcomes resulting from the threats to assets. For example, a series of terrorist threats could interrupt business and cause financial loss, or a tornado could cause severe property damage and casualties. In addition, loss of trust or confidence in community leadership, as a result of unanticipated threats, would be considered potential impacts.

Coombs's (2014) crisis typology can inform the way in which a CCC helps a community assess its prominent hazards and find ways to anticipate and manage them. For instance, a community might be located in a region that is susceptible to tornadoes (natural disasters). The community might have a manufacturing company that is controversial in the way in which it disposes of toxic waste, and therefore might be prone to malevolence, such as vandalism or rumors spread by disgruntled consumers or competing companies. Crisis managers and decisions makers need to be tuned in to the types of crises that affect their community the most and to the resources they have available to mitigate the impact of potential crises.

One simple but effective method of risk assessment is the SWOT analysis. The four letters represent an analytical method used to identify and classify internal (*strengths*, *weaknesses*) and external (*opportunities*, *threats*) that impact a community (Humphrey, 2005). Analysts collect and then convey or portray the information about these internal and external factors to enable leaders and policymakers in their planning, allocation of resources, and forecasting of events. The strengths and weaknesses relate to the presence or lack of a community's resources, assets, or capabilities. The opportunities and threats refer to the environment, usually factors over which the community has little or no control. Although this analysis was originally designed for planning a business venture or project (Humphrey, 2005), it can be effective in identifying threats to the community, as well as resources and capital that can be used to anticipate and to respond effectively. Figure 9.3 provides an example of a CCC SWOT analysis for a midsize town. The SWOT analysis does not just apply to crises, but to any issues that face a town, city, or region. The SWOT analysis is a more general way to capture the state of the community, and it can be preferable when there is a lack of individuals with crisis management expertise or experience. If a community already has an analysis such as this in place, then planners might prefer analytical methods that are more specific to risk assessment.

INTERNAL FACTORS	**Strengths** • Has a top-tier research university • Has a high-ranked health system • City planner has training and experience in crisis planning	**Weaknesses** • Sections of infrastructure damaged from recent tornado • City operating under a budget deficit • Lack of established communication system among fire dept., police dept., and medical facilities
EXTERNAL FACTORS	**Opportunities** • Two research universities in neighboring city • New media and social networking technologies are available	**Threats** • Geographic area vulnerable to severe tornado and thunderstorm patterns • Budget cuts by state legislature • Recent poisoning of water supply in neighboring city

Figure 9.3: Crisis-related SWOT Analysis. Sample SWOT Analysis of a Midsize City.

An effective risk assessment must be anticipatory in nature. More specifically, it needs to account for expectations, which are assumptions people make about objects, events, or the environment (Olaniran & Williams, 2001). Inaccurate assumptions about the threats most relevant to a community can leave that community unprepared for the magnitude of those threats and for how to moderate their impact. For example, a midsize city that does not foresee the threat of a terrorist attack might overlook the possibility of its water supply being polluted. According to Olaniran and Williams (2001), assumptions about a community's assets and the potential threats can provide decision makers with a starting point to assess risk. Finally, risk assessment involves the amount and quality of communication that occurs among sectors, between leadership and the public, and with the media.

Risk Management

Once risks have been assessed, a risk management plan can be developed, implemented, and periodically modified to best fit the changing demographics and needs of the community. Risks vary in their nature and type, probability of occurrence, and their potential impact. Risk management involves prioritization, which entails the allocation and coordination of resources needed to control the risks, as well as to monitor them and to minimize their impact (Hubbard, 2009).

Those charged with risk management can consult resources for assistance in developing a risk management plan, as well as learn about other communities that have put successful plans in place. One such resource can be found at Ready.gov, which provides an impact analysis model that helps organizations identify their own hazards and vulnerabilities (see the link at the end of this chapter). The CCC's role in risk management development mainly rests in advising agencies on who should be involved in development and implementation, as well as the problem-solving process that is required. Again, crisis managers should consider Coombs's (2014) typology when identifying the resources, information, people, and structures that would be most useful in mitigating potential disasters. For instance, if a local manufacturing plant produces byproducts that could be damaging to the water supply (mega damage), the CCC can facilitate a meeting among community leaders, industry leaders, local citizens, and environmental experts to discuss the potential harms from the mega damage and find ways to prevent or to handle the disaster.

Information and Databases

One of the most important tools available to a CCC is information, which should be available to CCC personnel and to individuals within the community. Information to have on hand can include facts about crises most relevant to the community, the general operations of the CCC and its mission and vision, community resources, the sectors instrumental in providing leadership during certain types of crises, and opportunities for education and training.

Brochures, fact sheets, web addresses, and emails can provide information about the CCC and about the organizations linked to it. In the event of a crisis, such information can be made readily available about the organizations affected or have even played a role. These materials are not specific to particular events or crises; they are generic in the information they contain about the organizations, and should be written and designed to enhance credibility, trust, and competence.

The CCC will be instrumental in collecting and warehousing information that can be accessed by various community sectors. Such information might include control banding and guidelines, community statistics (e.g., population demographics, economics, and weather and environmental hazards), lay publications, news articles, and other information considered relevant by the CCC liaison and by the agency partners. Also included would be training materials for personnel who work directly or indirectly with the CCC, as well as educational publications and articles available to the public.

Another important tool for the CCCs is a data asset management system (DAMS), which is an electronic storehouse containing content, images, photographs, infographics, videos, music, animations, and other assets that can be cataloged, stored, and retrieved by individuals putting together information for dissemination (van Niekerk, 2006). The creation of a DAMS requires the appropriate computer software and hardware to allow for downloading, data backup, naming and renaming, archiving, and exporting. The bylaws could stipulate what the CCC will catalog in the DAMS and who should have access to it. Preferably, the executive director, liaison, webmaster, and key support staff should be given access in order to manage the system regularly and efficiently. Other individuals might be granted permanent or temporary access depending on the type of information that is being used or disseminated at any point in time.

Signal Detection

Once risks have been assessed and prioritized, the CCC can enable community leaders and agencies to identify common signals of the risks affecting the population. Signal detection plans should be developed and implemented that allow each agency to monitor their environment for warning signs of potential disasters. For example, hospitals and healthcare providers can monitor the local environment for clues of public health threats, such as increased hospital admissions for a particular infection, dead or diseased livestock, and sudden increases in environmental pollutants.

The emergence of a signal might not necessarily foreshadow a widespread crisis; it might simply be a warning sign that the threat still remains at a low level. Because threats vary in their intensity and likelihood, decision makers within each sector should work with researchers and experts to generate a system of signals to monitor, with the signal being quantified to indicate different levels of the threat. For example, hospital liaisons working with the CCC could scan the environment for warning signs that their community is suffering from an influenza outbreak during a high-risk flu season. A low-level threat might be 10 flu-related hospital admissions per week. A signal indicating a mid-level threat will reveal higher numbers of admissions, and so on. When the signal reaches a certain level, then it would be time to alert other sectors to the threat, as well as link with the CCC to coordinate communication with the media and with the public.

A caveat needs to be made about interpreting signal levels or warning signs. Suppose a signal detection system consists of a scale of 1 to 5, with 5 indicating high risk and 1 indicating no risk. If the detection system reveals that a threat is at a level of 3.5 (moderate to moderately high), it might be tempting to assume that the 3.5 level does not signal a threat. However, vigilance of signals assumes that risks that have not yet reached a high level might be approaching that stage. Waiting until the signal indicates a high level is too rigid a response, and might be too late to ward off severe impacts of a crisis (Scholl, Williams, & Olaniran, 2005). On the contrary, a more flexible interpretation of the signal might compel a detector to take the potential threat seriously and even to take actions that would not have been possible under a more rigid approach to vigilance.

Another important aspect of signal detection is coordinating the communication of these signals across sectors and to the community at large. A signal system that is relatively simple and conceptual can be conveyed to the public easily. In the case of hazard banding, different threat levels are visually represented with horizontal bands of different colors, such as the National Terrorism Advisory System used by the U.S. Department of Homeland Security. Such systems can be an effective method for signal detectors to recognize the signs of threat quickly as well as to have a data visualization system that makes it easier to convey numerical information to the public.

The concept of banding is also useful to controllers who must execute certain controls depending on the level of harm or exposure to a threat. For example, the National Institute for Occupational Safety and Health (NIOSH, 2013) provides a control guide for chemical inhalation. NIOSH recommends that inhalations greater than .1 to 1 mg per cubic meter of dust signal harm during one single exposure, and the recommended control is local exhaust ventilation. Higher levels of inhalation will require more aggressive controls, such as closing the area or seeking medical advice. Through coordination by the CCC, representatives from the medical community might work with environmental engineers, law enforcement, and other first responders to determine the controls that are appropriate for each level of threat. In addition, the CCC can develop and disseminate information about these control banding systems to the public; such information might contain guidelines that could help community members determine which actions to take that are appropriate for the level of crisis threat. Information gathering and dissemination can be enhanced further when decision makers stay connected regularly to such organizations by signing up for e-newsletters, following blogs, and using social media such as Facebook and Twitter to get links to important information.

Communication Plans

The CCC needs to facilitate a plan that internal and external personnel (i.e., CCC internal staff, CCC liaison, sector liaisons, and organizational spokespersons) can use as a guide for communicating before, during, and after crisis events. Using Figure 9.3 as a guide, all personnel should have an updated list of people at all levels. The process should be centralized such that any person—whether CCC executive director or spokesperson of a company—should be able to initiate communication within the network should he or she detect a signal of a crisis or threat. While it might make sense for a company spokesperson to go first to his or her sector liaison, that spokesperson might also contact the CCC directly to facilitate a faster flow of information. Also, a spokesperson might not be able to contact his or her liaison, and might need to talk directly to someone at the CCC for advice on media relations if an interview or press conference is imminent.

Sector liaisons and spokespersons must be able to analyze their internal audiences to be able to understand and anticipate their needs during a crisis. According to Mathes and Stevenson (2011), there are the audiences in one's own group (project group or department), audiences in close proximity to one's group (other departments with whom you normally interact, often in adjacent or in management groups), audiences elsewhere in the organization (distant management, public relations, and those dissimilar to you in operation), and audiences outside the organization (i.e., other organizations). In addition to identifying names, titles, and roles, one also has to establish the purpose and content of each communication report or message for each audience.

With regard to communication outside the CCC, sector liaisons can organize members of their sectors to develop coordinated interorganizational communication plans, which enable members to have contact information of other members on hand and to get in touch with each other quickly. The CCC can centrally coordinate these efforts, partially with the help of the CCC liaison, who can advise on creating adaptable templates and utilizing message strategies. Most importantly, these strategies must be formulated with the external audiences in mind. In any community-level plan, one has to consider that audiences include employees and customers of an affected company; members of residential communities impacted; the chamber of commerce, distributors, retailers, and wholesalers; the media; the academic community (faculty, students, staff, and administrators); investors and shareholders; banking institutions; and local, state, and national government (Mathes & Stevenson, 2011). In short, these and other audiences affected by the likelihood of a crisis, or by an actual crisis, should be considered key stakeholders because they have the most to gain or lose by the proactive actions that anticipate a crisis.

Random practice drills should be conducted at least monthly that allow all CCC staff and external partners to practice their plans. Such a plan is intended to enable all individuals to send and receive information that would either meet a crisis threat head-on or respond quickly when a crisis does happen. Separate plans should be developed based on the risk assessment, which will identify the most salient threats, as well as the sectors and assets that are the greatest targets. A separate communication plan can be constructed for different threats, as not all threats have the same type and intensity of impact. A health-related crisis will require more involvement from the health sectors than a banking or financial crisis. Regardless of the plan, internal or external personnel who first identify the threat will be charged with sending out an alert that reaches the CCC executive director (who then contacts the board of directors), the CCC liaison and sector liaisons, and the organizational spokespersons most affected. Subsequent to the initial alerts, all affected personnel should report to the CCC in person or remotely to receive further instructions. In the meantime, the liaisons will be working with the spokespersons involved to craft or adapt a message to go out to the public and to the media. As the crisis progresses and as the media have received infor mation about the initial plan, the CCC liaison facilitates continued contact among the sector liaisons and the spokespersons to address issues as they arise. This follow-up process is overseen by the executive director and by the board of directors.

When spokespersons are called on to communicate with the media and to disseminate information to the public, it will help them to rely on standard messages that can be adapted to particular audiences and situations. The importance of standard messages—the kind that can be adapted to fit the crisis and occasion they intend to serve—resides in their easy and quick access, especially if they are stored in DAMS. A rapid response is one of the most important aspects to effective crisis management and to minimize or alleviate damage to the organization and harm to the community (Olaniran, Scholl, Williams, & Boyer, 2012). Although it might take some time to alter a template to fit the situation at hand, having that template in place can save crucial time otherwise needed to control certain events (e.g., keeping information from prematurely leaking to the media and preventing certain harm from reaching vulnerable neighborhoods).

One standard message to have ready is a brief description of the crisis (or the threat of one) affecting the community. When designing a prepared statement (e.g., one given during a press conference), consider the who, what, where, when, why, and how as much as possible. A statement template might look like this:

> A [event] at [location of crisis or tragedy] involving [known victims and/or perpetrator(s)] occurred ["today" or date] at [time of day]. The incident is under investigation and more information will be provided as it is verified and made available.

In most cases, simply filling in the blanks of this template will not suffice, but it gives a starting point. Additional information, such as when the next news release will occur, when the next news conference will take place, or what damage/casualties have occurred, might be crucial to the initial message.

Press releases can be planned in such a way as to serve specific audiences with the most appropriate information. Wilson (1997) provides a press release outlined in Figure 9.4.

Press Release
Copy Outline

Specific targeted public(s):

Secondary public(s):

Intervening public(s) (if any):

Specific tactics to target intervening public(s):

Action desired from public(s):

How that action ties into public's self-interest:

Main message to be conveyed:

Secondary message or objectives:

Proposed headline for this news release:

Proposed lead for this news release:

Copy points (as many points and as much specific information as you need to communicate to your public to support your message):
1.
2.
3.
4.
(Use all copy points necessary to completely detail the information.)

Specific media to receive release:

Delivery method and follow-up activities with media (if needed):

Figure 9.4: Copy Outline for a Press Release.

Information that follows an initial statement, and communication intended to repair damage, requires some strategic planning. In these cases, an adaptable template might not be practical, but keeping some standard strategies in mind will make it easier for organizations to reestablish credibility and to restore public trust. One tactic has to do with presenting facts. Make sure the affected organizations have verified those facts and that they come from reliable sources. Broadcasting "facts" before they are verified will only encourage scrutiny and counterarguments from the media and from the public.

When verifiable facts are not available, but the affected organization(s) need to reach out to the public, the CCC can coach a spokesperson to communicate the organization's concern. This does not necessarily reflect a failing, but actually reflects a step in the right direction. An organization that does not reach out to the public risks losing credibility. The CCC liaison must remind the organization that the public will want information. In this case, the spokesperson must assure the audience that key facts will be shared as soon as they are verified and can be made available.

Another important tactic that spokespeople can learn is to anticipate tough questions and to prepare competent answers (PR Fuel, 2014). The CCC liaison can help the spokesperson identify potential questions and provide coaching on the best responses. Tough questions, particularly from journalists, will try to get at the who, what, where, when, why, and how. Questions will also likely center on what happened, what was the cause, and what the event means.

In addition, there are tactics that can help spokespeople prepare for tough questions from the media (PR Fuel, 2014). One is to not answer a question that lacks validity, is too personal, or is inappropriate. A spokesperson has the right to attack or avoid the questions. Also, refer the reporter to someone else if the question doesn't relate to the spokesperson's expertise or area. Additionally, the spokesperson can offer to find the answer and to get back to the reporter. Also, if the spokesperson already answered the question, that question should not be answered again if repeated. For example, if asked, "Do you know who or what caused the fire?" One might answer, "We do not yet know the cause, but are conducting an investigation into the matter." The reporter might come back with the same question reworded: "We have indications that some of your own workers could have caused the fire. Can you comment?" The reply could be, "We have already addressed that question."

To minimize the chance of being asked a tough question, keep the statement simple. This is not to be deceptive, but to prevent the message from seeming complicated and thus alienating some members of the audience. For example, if a crisis involves a piece of equipment, one might provide a model or illustration of that equipment for illustrative purposes. Related to keeping it simple, the message should have one and only one aim. Vinci (2011) suggests seeing the message as a missile aimed at one target, not a torrent of shotgun pellets that are scattered erratically.

Although the advice has been to keep a message simple, this does not mean that spokespeople should focus on giving less information. On the contrary, the more information spokespersons provide, the less likely the public or media will seek out that information for themselves, which can often lead to misinformation or confusion (Scholl et al., 2005). This also means defining terms that might have different meanings across industries, cultures, or audiences.

There might be some information important to the public that the affected organization has to disseminate with CCC assistance. For instance, a crisis that poses certain health risks might not get adequately reported in some newspaper outlets. This was found by Cohen, Vijaykumar, Wray, and Karamehic-Muratovic (2008) in their examination of newspaper reports of health risks following Hurricane Katrina. Not only might media sources provide little information, but they also might lack information about expert sources, whom organizations might consult on their own in getting crucial information to the public. Therefore, particularly the CCC liaison and the sector liaisons can work together to find experts from the appropriate sectors, who can help craft more complete and accurate messages that the spokespersons can provide to the public.

Education and Training

Another important function of the CCC is to provide continuing education and training for its staff, board of directors, and its partnering agencies. The CCC assumes that while internal staff and board members are capable of handling a crisis from their own professional perspective, they might not be equipped to handle various communication issues inherent in crisis management and in damage control. Thus, the CCC needs to provide continued training in these areas: being comfortable in front of TV cameras, on radio, and with reporters; staying on point with topics; identifying and articulating key main points; speaking to the public without using jargon; being knowledgeable about the crisis and their organization's role; conveying competence and charisma; obtaining and maintaining public trust and respect; and remaining calm in stressful situations.

There is a distinction between education and training (Beebe, Mottet, & Roach, 2013). Education involves disseminating knowledge, facts, and other information for people to use. Being educated on a topic means having exposure to many ways to achieve a goal and obtaining information that is linked to a broad topic rather than to a specific task. Training involves acquiring and developing skills for a specific job or task, and attaining a specific skill level that is appropriate for the task.

What this means for the CCC is that some topics are appropriate for educating the public, such as providing information on the types of disasters that most affect a community. Training, however, means that there are specific skill sets that enable community members and decision makers to prepare for, and react to, a crisis (Beebe et al., 2013). The CCC can organize training programs and materials on information dissemination for spokespersons, organizational leaders, and people who speak to the media. For instance, training sessions can teach individuals best practices for communicating honestly while engaging damage control and maintaining message simplicity. Public speaking is another valuable skill, and such training can be facilitated by CCC personnel or by local experts partnering with the CCC. Training plans should touch on audience analysis, nonverbal skills, delivery styles, language skills, and media interviews, as well as other skills of interest to trainees.

Another important aspect of training involves the risk and crisis management process, which is covered in Chapter 10. This process should be familiar to CCC leaders and personnel, as well as to individuals representing partner agencies. It would be advantageous for the CCC to conduct these types of training at regular intervals throughout the year. What follows are the specific components of the risk and crisis management process that can be adapted to the needs of the community.

Conclusion

The purpose of this chapter has been to provide a procedural and operational framework for local crisis communication centers. This framework includes clear mission and vision statements, key personnel, such as a board of directors, an executive director, support staff, and a CCC liaison who relates directly to representatives of community sectors. The chapter then provided guidelines for general operations of the CCC, which includes bylaws and a strategic plan. Finally, a crisis communication management plan was outlined, which local CCCs can adapt to their own community-level needs. This framework promotes the inclusion of a communication-based risk assessment, a risk management strategy, a system of signal detection, information warehousing and databases, communication plans, and continued education and training for CCC personnel, liaisons, and community stakeholders. Having a CCC operational framework not only enhances the communication necessary for effective crisis management, but it also can provide a model for other communities to use.

The following are resources that can guide the development of CCC procedures and operations:

- Community Tool Box: Proclaiming Your Dream. Developing Mission and Vision Statements. http://ctb.ku.edu/en/table-of-contents/structure/strategic-planning/vision-mission-statements/main
- Ready.gov. Risk Assessment: http://www.ready.gov/risk-assessment

References

Abrams, R. (2003). *The successful business plan: Secrets and strategies*. Palo Alto, CA: Running 'R' Media.

Armstrong, J. S. (1986). The value of formal planning for strategic decisions: A reply. *Strategic Management Journal*, *7*, 183–85.

Beebe, S. A., Mottet, T. P., & Roach, K. D. (2013). *Training and development: Communicating or success* (2nd ed.). Boston: Pearson Education.

Cohen, E. L., Vijaykumar, S., Wray, R., & Karamehic-Muratovic, A. (2008). The minimization of public health risks in newspapers after Hurricane Katrina. *Communication Research Reports*, *25*(4), 266–81.

Coombs, W. T. (2014). *Ongoing crisis communication* (4th ed.). Thousand Oaks, CA: Sage.

Federal Emergency Management Agency (2012). Risk assessment. *Ready.gov*. Retrieved from http://www.ready.gov/risk-assessment

Foundation Center. (2013). Establishing a nonprofit organization. *Foundation Center.org*. Retrieved from http://foundationcenter.org/getstarted/tutorials/establish/board_dev.html

Fram, E. H., & Brown, V. (1995). *Policy vs. paper clips: Selling the corporate model to your nonprofit board* (2nd ed.). Milwaukee, WI: Families International.

Heath, R. L. (2001). The dynamics of change in public relations practice. In R. L. Heath (Ed.), *Handbook of public relations* (pp. 183–88). Thousand Oaks, CA: Sage.

Hill, C. W. L., & Jones, G. R. (2008) *Strategic management: An integrated approach* (8th ed.). Mason, OH: Southwestern Educational Publishing.

Hom, E. J. (2013, October 23). What is a vision statement? *Business News Daily*. Retrieved from http://www.businessnewsdaily.com/3882-vision-statement.html

Hubbard, D. (2009). *The failure of risk management: Why it's broken and how to fix it*. New York: Wiley & Sons.

Humphrey, A. (December 2005). SWOT analysis for management consulting. *SRI Alumni Newsletter*, 7–8. Stanford University: SRI International.

Kirkpatrick, S. A., Wofford, J. C., & Baum, J. R. (2002). Measuring motive imagery contained in the vision statement. *The Leadership Quarterly*, *13*(2), 139–159.

Lipton, M. (1996). Demystifying the development of an organizational vision. *Sloan Management Review*, *37*(4), 83–92.

Mathes, J. C., & Stevenson, D. W. (2011). Audience analysis: The problem and the solution. In K. J. Harty (Ed.), *Strategies for business and technical writing* (7th ed.) (pp. 168–83). Boston: Longman.

National Institute for Occupational Safety and Health. (2013). Control banding: Frequently asked questions. Retrieved from http://www.cdc.gov/niosh/topics/ctrlbanding/ctrlbandingfaq.html#10

Olaniran, B. A., Scholl, J. C., Williams, D. E., & Boyer, L. (2012). Johnson and Johnson phantom recall: A fall from grace or a re-visit of the ghost of the past. *Public Relations Review*, *38*, 153–55, doi:10.1016/j.pubrev.2011.08.001.

Olaniran, B. A., & Williams, D. E. (2001). Anticipatory model of crisis management: A vigilant response to technological crises. In R. L. Heath (Ed.), *Handbook of public relations* (pp. 487–500). Thousand Oaks, CA: Sage.

PR Fuel. (2014). 5 tactics for dealing with tough questions from the media. *PR Fuel: PR news, views, and stews*. http://www.ereleases.com/prfuel/5-tactics-for-dealing-with-tough-questions-from-the-media/

Sapriel, C. (2003). Effective crisis management: Tools and best practice for the new millennium. *Journal of Communication Management*, *7*(4), 348–55.

Scholl, J. C., Heuman, A. N., Olaniran, B., & Williams, D. E. (2007). Communicating during terrorist threats and acts through crisis communication centers (pp. 399–409). In S. C. Duhe (Ed.), *New media and public relations*. New York: Peter Lang.

Scholl, J. C., Williams, D. E., & Olaniran, B. A. (2005). Preparing for terrorism: A rationale for the crisis communication center. In D. O'Hair, R. Heath, & G. Ledlow (Eds.), *Community preparedness and response to terrorism: Communication and the media* (pp. 243–68). Westport, CT: Praeger.

Stephens, J. L. (2000). *Bylaws: Writing, amending, revising* (2nd ed.). Clearwater, FL: Frederick Publishers.

van Niekerk, A. J. (2006). *The strategic management of media assets: A methodological approach.* Allied Academies, New Orleans Congress.

Vinci, V. (2011). Ten report writing pitfalls: How to avoid them. In K. J. Harty (Ed.), *Strategies for business and technical writing* (7th ed.) (pp. 203–9). Boston: Longman.

Wilson, L. J. (1997). *Strategic program planning for effective public relations campaigns* (2nd ed.). Dubuque, IA: Kendall/Hunt.

Wilson, L. J. (2001). Extending strategic planning to communication tactics. In R. L. Heath (Ed.), *Handbook of public relations* (pp. 215–22). Thousand Oaks, CA: Sage.

Case Study

FEMA Strategic Foresight Initiative

Source: http://www.fema.gov/media-library-data/1422989258096-6aeb946a17ccb021d5d994fa1b3218b5/2_SFI_Case_Studies.pdf

In an effort to learn how emergency management practitioners around the country are using foresight, the FEMA Strategic Foresight Initiative (SFI) Team sat down with Sean Johnson, plans officer in the North Dakota Department of Emergency Services, to get a sense of how he is using SFI ideas and materials. We also asked him for some recommendations on how to make foresight useful to emergency management (EM) practitioners.

What is the mission of your organization? Our mission is to conduct planning, coordination, communications, and operations for the safety and security of all North Dakota citizens. We coordinate emergency management activities across the five National Preparedness System mission areas for the state's 53 counties and two tribal nations.

What is your role or area of specialization? I am the plans Chief in the Operations and Planning Section. We develop the State Emergency Management Plan and supporting documents, conduct the Threat and Hazard Identification and Risk Assessment process, and manage EM information and intelligence, including state warnings, the Integrated Public Alert and Warning System (IPAWS), and Amber alerts.

How did you learn of SFI?
I first learned of SFI through a public/private partnership conference in Colorado Springs in 2012, and then again at the National Preparedness Conference in 2012.

How do you use SFI in your work?
One part of my role is instructing emergency management officials from across the state on emergency operations and tactical level planning. This used to be a two-day course, but I've expanded it to three days to use one full day to cover SFI since it provides a useful overview of strategic challenges the EM profession will face over the next 15–20 years. It also lays out a strong framework for how I want my students to think for the rest of the class. While the course doesn't focus on the strategic level, the SFI piece gives them "the road ahead" and an idea of strategic planning and thinking.

Walk us through what we would experience if we took one of your courses.
We start with an SFI presentation. I used the presentation from the National Preparedness Conference in 2012, but I've added information specific to our state. We spend the morning talking through SFI concepts around drivers of change and strategic needs. In the afternoon we break into groups to conduct a Strengths Weakness Opportunities and Threats (SWOT) analysis. I ask participants to think about what they need to account for in their plans, which

sets the stage for the next two days. The same themes keep playing out for the duration of the course.

Are you familiar with the resources available on the SFI website? Do you use them? If so, which ones?
I use the driver papers to spur thought during my classes. In one discussion of the driver paper on demographic shifts, we realized that North Dakota aligns with the nation on some trends, but bucks others. For example, our infrastructure is wearing out in North Dakota due to industry truck traffic, but we also have a $3 billion budget surplus and a labor shortage. National trends point to people moving to the coast, but North Dakota's population is growing due to the energy industry.

How would you recommend others get involved or use SFI?
The SFI drivers impact the whole community—they are a catalyst to bring everyone together. SFI is the flag that everyone can rally around. Even just including the drivers in planning discussions or sharing the SFI reports could be a good way to get people to think strategically and then apply that forward thinking in their emergency and operations planning.

Discussion Questions:

1. How well does the mission of the North Dakota Department of Emergency Services adhere to the characteristics of an effective mission statement as outlined in this chapter?
2. What other kinds of organizations and agencies should the North Dakota Department of Emergency Services partner with to prepare for various crisis events and why?
3. In addition to the SFI website, what other types of resources would be useful to the department, and why?

The views expressed in this case study do not necessarily represent the official views of *the* United States, the Department of Homeland Security, or the Federal Emergency Management Agency.

· 10 ·

IMPLICATIONS FOR PRACTITIONERS AND COMMUNITY SECTORS

Chapter Highlights

- The effectiveness of the CCC network
- Pre-crisis measures should be proactive
- Communication during the crisis phase
- CCC communication during the post-crisis phase

The purpose of this chapter is to provide guidelines for sectors, liaisons, and spokespersons who partner with the CCC. Common sectors, such as public relations, medicine and healthcare, police and fire, city council, commerce and retail, and education, all have their unique needs when it comes to preparing for, and minimizing the effects of, crises. Additionally, these sectors and the organizations and businesses contained within them need to operate together in such a way as to centralize their communication network.

This chapter addresses the practical implications of the CCC as they apply to three major phases—pre-crisis, during the crisis, and post-crisis. Although this book has emphasized the importance of anticipatory crisis management, the CCC, in cooperation with various community sectors, can take practical steps during all three phases to engage in forecasting events and in implementing effective crisis communication. Discussed in each of these phases are the unique needs of prevalent community sectors and the roles their spokespersons and liaisons (representatives) play in all three crisis phases. Also, guidelines will be offered to assist in the creation of specific crisis plans that can be utilized in partnership with the local CCC. To this end, the CCC network of personnel outlined in Chapter 9 will be reviewed and its functions will be outlined. In this chapter, the CCC network refers to CCC internal staff, CCC liaisons, sector liaisons, and spokespersons, and anyone else who partners with the CCC.

Review of the CCC Network of Personnel

The CCC network outlined in Chapter 9 contains members inside and outside the CCC facility. The network relies on the coordination and collaboration among key community members to make the CCC operate effectively. Internal to the CCC is the executive director and the support staff. With the crucial assistance of the support staff, the executive director can oversee CCC operations as well as coordinate and provide direction for the CCC network and other external personnel. The executive director's decisions are informed by the board of directors, who consist of leaders who represent various sectors of the community.

External to the CCC are the CCC liaison (or liaison team), the sector liaisons, and the organizational spokespersons. The CCC liaison is the link between the CCC and the participating organizations and agencies who receive assistance from the CCC. Through their regular contact with the CCC and with the CCC liaison, the sector liaisons help appoint and communicate with spokespersons for the agencies, businesses, and organizations who want to partner with the CCC. These spokespersons are typically the faces of their organizations during times of crisis, and they need the support of the CCC and its external partners in communicating effectively within their organizations and to the community at large.

The executive director, board of directors, support staff, and other key individuals in the community need to discuss how the CCC should operate and be made relevant and visible to the community. The CCC liaison should maintain a good working relationship with the sector liaisons and communicate regularly with spokespersons of partnering businesses, agencies, and organizations. Moreover, these partners need to map a strategy for handling the stages of various crises that could threaten their community.

The Pre-crisis Phase

What follows is a discussion of each phase, the challenges and implications of preparing for each phase, and ways in which the CCC network can be prepared. Exploring the practical implications of each crisis phase is warranted because the phase of crisis development is among the factors that can vary and thus is likely to affect how communicators create and select messages and respond to individual crises (Seeger, 2002). For example, in the pre-crisis stages, it is wise to take a more anticipatory approach by foreseeing the potential impacts and thus taking the steps necessary to reduce or to minimize harm. However, as the crisis materializes and starts to impact certain community functions, decision makers and spokespersons must adapt to the crisis as it unfolds. Crisis phases or life cycles are important to consider because crises are dynamic events. As the crisis moves through its stages, communication needs change, as well as the strategies that are used to understand and respond to the crisis (Massey, 2001).

Furthermore, this section addresses these three phases with respect to different sectors or types of organizations that comprise the community. According to Millar (2004), certain industries tend to be prone to crises—finance and insurance; information systems and telecommunications; scientific, professional, and technical services (e.g., aerospace and pharmaceuticals); oil and gas mining; transportation and warehousing (e.g., air transportation); and manufacturing (e.g., motor vehicles). The identification of these vulnerable groups is based largely on similarities in function, operations, and overall organizational needs (Glynn, 2000).

Seeger (2002) points out that organizational type also can affect message creation and crisis response. While the type of organization or sector (e.g., business, medical, and nonprofit) has not often been considered a variable in crisis communication research, it can influence how organizational members respond to crises (Massey, 2001). The routines, functions, and the identity of an organization can influence how well individuals respond to a crisis, understand it, and communicate important information about unfolding events to the public (Ginzel, Kramer, & Sutton, 1993). For instance, if a for-profit business has a culture that values the bottom line at the expense of public safety and transparency, it might respond to a crisis by hiding information that reveals inefficiency or misconduct as causes of recent crisis events. Such a response would be an attempt to preserve the company's image, but it does not uphold transparency or the right of the public to know whether the misconduct threatened public safety or health. Such was the case with Johnson & Johnson when they implemented a "phantom" recall of Children's Motrin (Olaniran, Scholl, Williams, & Boyer, 2012), a product that sometimes failed to dissolve properly when taken by mouth. Instead of publicly recalling the faulty product, Johnson & Johnson hired "consultants" to buy the product off store shelves in secret. While the company maintains that no consumers were hurt by the defective product, preserving their corporate image and profits apparently outweighed the value of keeping the public informed and thus maintaining transparency.

Communication during the pre-crisis stage is typically marked by proactive approaches (Gonzales-Herrero & Pratt, 1996). Proactive communication strategies serve to reduce the likelihood of crises as well as to prepare those markets and populations that could be affected (Heath & Millar, 2004). One proactive strategy is issues management—an environmental scan that identifies potential problems before they become real (Bowen, 2012). Issues management is anticipatory in nature and is contrary to the approach of managing a crisis after things collapse (Pauchant & Mitroff, 1992).

One simple but important tactic of issues management is to brainstorm all of the crises that could occur within a community, regardless of how rare they might appear to be. This might prove to be difficult for individuals who are trying to think of events that have a high likelihood of occurrence. On the contrary, a crisis, by its nature, is not something that is likely. However, when a crisis does happen, its impacts can be devastating and long-lasting. This is why crisis managers should be directed to think of crises that are not necessarily likely, but that could have a significant impact. Crisis managers are also encouraged to identify issues that naturally lend themselves to crises, such as labor strikes, economic downturns, and unsafe working conditions (Bowen, 2012). CCC liaisons, sector liaisons, spokespersons, and organizational leaders should generate an inventory of possible crises and work together to determine which crises cross sector boundaries. It is especially important that these leaders do not work in isolation, but together to identify relevant crisis scenarios.

In addition to the issues management process outlined in Chapter 8, provided here are Jaques's (2012) areas of pre-crisis issues management, one of which is "proactively addressing underlying systemic causes of potential crisis" (p. 23). Identifying the potential for a crisis might not be as difficult as understanding the possible causes. This is why cross-sector collaboration within the CCC network is essential. For instance, a disease outbreak might have multiple causes. Medical practitioners might identify poor hygiene practices, the social services sector could point to poverty, and the business community might attribute an outbreak to noncompliance with food-handling mandates. When different sector liaisons and spokespersons discuss the same type of crisis, they are likely to identify more causes together than they would if they worked individually. Furthermore, a more complete picture of the causes could shed light on the kinds of feasible solutions and bring the realization that the same solution might alleviate more than one cause.

The second issues management area that Jaques (2012) identifies is "establishing effective signal detection mechanisms" (p. 23). The disconnect between information and action is at the heart of system failure. If managers and decision makers do not have accurate information about the system and the environment in which it exists, they are unable to act in timely and appropriate ways when the system malfunctions.

Furthermore, many crises are marked by clear warning signals that are often ignored (Jaques, 2012). For example, a hospital's emergency room experiences a slight increase in a common set of physical symptoms that indicate meningitis. Moreover, most of these patients report eating at the same restaurant within the previous week. The ER physician who studies the cases notices the increase in admissions is only 4% in the last two weeks; this figure is lower than the 5% the hospital needs to see in order to take action. If the physician decides to ignore the warning signs because of strict adherence to the threshold of acceptance, this could lead to disaster. On the other hand, bringing attention to this marginal increase could allow public health officials and community leaders to take action before the crisis gets out of hand.

To address signal detection, sector liaisons and spokespersons across sectors should compare their guidelines and signal detection protocols. It is also important that they educate each other on their own industry standards with regard to such factors as chemical exposure levels, hazard controls, and safety measures. Part of this discussion should also involve generating recommendations and suggestions for changes in current standards, all of which should be shared with the CCC and across the CCC network.

Issues management also includes "properly identifying stakeholders and their perspectives" (Jaques, 2012, p. 23). Sector liaisons and spokespersons are in the best position to represent their industry perspectives and interests. Identifying these interests is one of the best places to start in terms of taking a complete inventory of the community groups and segments potentially affected by a crisis. For example, sector liaisons from the medical and commercial food sectors might realize that many of their stakeholder interests intersect, especially when food poisoning can threaten both a manufacturer's reputation and a hospital's surge capacity. In addition, not all stakeholders might live or work in the community (e.g., merchants from out of town, or in neighboring cities who share natural resources).

Finally, issues management should involve "learning and unlearning on an ongoing basis" (Jaques, 2012, p. 23). This means routinely identifying harmful or dysfunctional practices and bad habits, as well as cultivating and promoting best practices. Jaques asserts that learning and unlearning take place when crisis managers document what occurred in all three crisis stages, noting successes and mishaps, and making recommendations for change. These recommendations and other insights must be communicated upward—to CEOs, to city council members, and to others who have decision-making authority. Overall, the process of learning and unlearning provides the CCC with an opportunity to establish itself as a center for best practices. The findings documented and the lessons learned from issues management and from post-crisis assessments can allow the CCC to compile a dynamic store of information, principles, case studies, and protocols that can be shared with experts and lay people not just inside the community, but with other communities who want a model to develop their own CCC.

The network of personnel in collaboration with the CCC can, and should, participate in issues management. Under the advisement of the board of directors, the CCC executive director should work with sector liaisons and spokespersons to coordinate community-wide issues management plans. Once the issues management areas have been identified and discussed by the liaisons and spokespersons, the findings should be shared with CCC personnel who can keep this crucial information on file. A data asset management system (DAMS) (see Chapter 9) is a good tool for keeping information on file for easy reference.

Another aspect of pre-crisis planning for the CCC network is the preparation of crisis materials that are readily available when disaster strikes (Gonzales-Herrero & Pratt, 1996). These efforts include developing position statements that can be adapted for particular events and for stakeholders. These statements would be used to communicate immediately with the public after the initial impact of a crisis or when the threat of a crisis has been established. Organizational spokespersons can work with their sector and with CCC liaisons to develop a standard statement as well as determine which message components are adaptable. Other precrafted communication tools include responses to frequently asked questions and standard templates with which to communicate with stakeholders (Gonzales-Herrero & Pratt, 1996).

Audience Issues

Standard crisis messages should exist not only for press conferences and interviews but also for questions individual community members might have. If adequately staffed with paid or volunteer workers, the CCC can serve as the hub for questions about an issue, threat, or existing crisis. The key is for the public to be informed of this option; such an announcement should be made regularly through local media outlets and on the CCC website. Questions from the general public might ask about where donations might be made, what should be done to protect oneself, and how one can reduce risk or exposure to the threat.

CCC staffers who take the inquiries should have a list of responses to common questions, but if they are not able to answer a question, they should forward it to an appropriate liaison or expert as soon as possible. Liaisons and spokespersons can identity subject matter experts (SMEs) ahead of time to consult with such questions. SMEs can take inquiries on a regular basis, but they should be on call during severe threats or actual crisis events. To streamline the process further, SMEs can be assigned to certain types of inquiries. For example, a physician assigned as an SME would receive health-related inquiries, an environmental toxicologist at the local university might take the inquiry related to watershed or soil contamination, and an industrial hygienist would advise on chemical exposure levels in workplaces.

To ensure that no inquiries are overlooked, the CCC should maintain a log that documents phone calls, e-mails, and personal inquiries from community members, officials, media personnel, and other individuals. Inquiry logs should contain the following elements: date of the inquiry, name and title (if applicable) of the person asking the question, question(s), telephone number(s) and e-mail address of the person inquiring, SME assigned to the inquiry, primary person who should receive the response (it might not be the person making the inquiry), and additional notes. A time frame should be established for when inquirers should receive a response; this time frame might vary depending on whether or not the question is asked during a crisis. A technical communication expert at the CCC or one recruited from outside, should conduct a communication review of the response to ensure accuracy, clear and basic language, and proper use of tone. Once the response has undergone a communication review, it then should be forwarded to the inquirer.

Times of tranquility are good opportunities to use social media channels, such as Facebook or Twitter, to establish and to maintain a connection with the community and to build trust and credibility. Increasingly more people are tapping into their smartphone and tablet to check their social media feeds to see what is trending. Smartphones have surpassed computers as the device of choice for using online search engines, such as Google (Sterling, 2015). The increased reliance on mobile devices means that CCCs should use social media to help audiences stay in touch with what the CCC is doing, and what conditions are emerging that could make certain crises more likely. More specifically, posts on Facebook and Twitter should be anticipatory in nature and remind users of upcoming events or conditions, such as winter weather or the flu. Messages should contain alerts and be proactive. For example, the phrase, "What you need to know about the Zika virus: Is our community at risk?" should be accompanied by a hyperlink to a website with more information, ideally the CCC's website.

Relationship with the Media

Another key communication issue for the CCC network is to establish and maintain a positive working relationship with the media. The only way that some agencies and organizations are linked to the public is through the media (Scholl, Williams, & Olaniran, 2005). A good relationship can ensure that the media serve as appropriate agents between the community and the CCC network. Such a linkage can facilitate delivery of important information when the community needs it the most.

Ideally, the CCC would serve as the hub for spokespersons and for other individuals who have the responsibility of talking to the media. In other words, messages to and from the media should be channeled through the CCC. For instance, if a spokesperson schedules a press conference, the CCC can serve as the location if there are adequate resources and space. Spokespersons can also request that preplanned interviews take place at the CCC, which can allow the spokesperson to receive consultation or coaching beforehand. If the interview or press conference cannot take place at the CCC, its staff should still be notified and remain on call for short-term communication coaching before and after the event. An alternative location for press conferences and interviews would be the spokesperson's organization or headquarters. However, the CCC is more ideal because it takes the media attention away from the involved organization, and prevents the media from being "in the middle of the action."

Certain guidelines for talking to the media should be part of regular training for all internal CCC personnel, liaisons, spokespersons, and anyone who is likely to be contacted by reporters. Key CCC personnel should be trained regularly on interviewing protocol. For instance, the personnel should be instructed to remain cool and confident, to answer only the questions they are asked and to do so succinctly, to know when to say, "I don't know," and to avoid common interview traps, such as questions that ask for conjecture or requests to speak off the record (Polevoi, 2013). Figure 10.1 offers guidelines for specific interview contexts.

Broad Interviews	TV Talk Shows	Newspaper Interviews
—Prepare primary talking points —Anticipate questions and prepare some answers —Discuss controversial topics early. —Find out who will be interviewing you if possible. —Find out how much time you will have during the interview.	—Use your regular voice and note how it comes across the microphone. —If makeup is offered, use it. —Sit in your chair with your back erect and lean forward slightly. —Use moderate hand gestures that are purposeful. —Smile and be friendly. —Keep your eyes on the interviewer, not the camera.	—Get the interview topics or questions in advance. —Establish a time limit before the interview takes place. —Avoid jargon or professional terms that most readers won't likely understand.

Figure 10.1: Guidelines for Types of Interviews. Adapted from American Psychological Association (2014) and Polevoi (2013)

During the Crisis

The crisis phase requires effective communicative efforts to address the crisis that has already occurred (Gonzales-Herrero & Pratt, 1996). These efforts include appropriately dealing with the media and making available crisis materials, such as position statements, responses to frequently asked questions, and messages, to stakeholders. The function of communication during a crisis is to target the right audiences, to gain support from other parties or community sectors, to maintain proper internal communications, and to control the spread of rumors.

Communicating the Central Message

One of the most important functions of the CCC at the unfolding of a crisis is to coordinate the central message communicated to the public. As the disaster progresses, spokespersons and liaisons from the affected sectors should meet with each other and with CCC communication experts. At this meeting, they need to determine their common understanding of the crisis, verify that their understanding is accurate, determine the best response to the crisis, and develop a plan of communicating with the public through the proper media channels. The coherent message that is communicated to the public depends on comprehension among group members of the nature, cause, and impacts of the current crisis. These factors will determine to a great extent who should be involved in the crisis management strategy facilitated by the CCC.

Once a common understanding of the crisis is reached, a crucial step is to decide on the appropriate response. Chapter 9 provided recommendations for a template that could be used to deliver a central message to the public and to the media. It would be helpful to have a core message prepared for each type of crisis anticipated, a message that could be adapted to fit the particular situation. Once liaisons and spokespersons work together to identify a potential crisis, they can seek assistance from the CCC in crafting these template messages (see Chapter 9 for a sample template). Regardless of how the message is adapted, it should convey the nature and type of crisis, its causes, who is affected, and what is being done to remedy the situation. This core message can then be adapted for a press release, press conference, public service announcement, content for the CCC website, or any other format that needs to be used.

In addition to the message, decisions should be made about who should deliver the response. Spokespersons (or people at the top leadership positions) of the organizations affected should be the ones to address the public and the media. It is possible that a crisis will affect several sectors. As in the case study at the end of this chapter, a meningitis threat affects the medical, business, and other sectors. Therefore, it is reasonable that spokespersons from each organization or agency affected will address the public with their specific messages. For instance, the local or county health department might have information on what kinds of treatments are available for people who are exposed to the illness. A hospital would inform the public about the number of cases treated and the symptoms to watch for. The restaurant from where the infection originated might tell the community what it is doing to identify the source(s) of the infection, what stores in the chain have been affected, and for how long the restaurant will be closed. In addition to the communication-related training these individuals should have received from the CCC, these spokespersons would ideally receive consultation from the CCC prior to communicating publicly about the current crisis. They would also be given access to the CCC for press conferences, interviews, and meetings with other crisis managers (if resources and space allow).

Use of the CCC's social media channels can be valuable in reinforcing the central message formulated by the CCC. Whether it's a blog post or a tweet, the main message should reinforce—not contradict or deviate from—the one that is communicated via press conferences and more traditional media. The tone of the social media messages is important. It should reflect a manner that alleviates the more counterproductive reactions that might occur. Also, as during the pre-crisis, messages should be proactive and direct users to more central sources of information, such as the CCC website.

Audience Issues

One of the keys to crafting an effective crisis message is understanding and anticipating how the community will likely respond to the situation and to the message. Citizens typically exhibit patterns of responses (Perry & Lindell, 2003). The initial responses, depending on the disaster, can include panic, shock, and passivity. It is rare that people will engage in a panic flight. Rather, they will often find ways to act in what they perceive to be their best interest, despite how little they might know about the situation. Not only do citizens engage in self-care, but they also use the information and resources available to them to help their neighbors and those in need.

Liaisons and spokespersons should be knowledgeable and competent enough to understand and anticipate the responses within their communities, and this knowledge should be shared throughout the CCC network. In addition, SMEs recruited to collaborate with the CCC should be consulted to provide insights into community members' physical, emotional, and psychological needs during a crisis. Such experts should have an understanding of the demographic and psychographic characteristics of the community and possess the skills necessary to communicate with diverse audiences.

Relationship with the Media

Media coverage of crises is unavoidable (Moore, 2004). According to Seeger (2002, 2006), an understanding of the nature of the crisis and the channels used to convey information are important in enhancing the success of an organization's crisis communication efforts. In negotiating a successful relationship with the media, a delicate balance must be maintained between the press's right to be on the scene and the organization's right to minimize disclosing damaging information (Scholl et al, 2005). Reporters are there to get the story out to the public, but the CCC must be able to work with leaders and spokespersons in managing the flow of information. The public's desire for information must be weighed against the organizational boundaries that need to be maintained. Too much information could overwhelm an audience. Claims that are not verified could lead to poor decisions made by those who receive the faulty information. Perhaps the best people to judge what information is helpful versus hurtful are the liaisons and spokespersons who represent the interests of the stakeholders and the affected organizations.

Another effective use of the media pertains to knowing which channels should be used (Scholl et al., 2005). For instance, television and live stream on the Internet can be used to convey late-breaking news and urgent calls to action. In contrast, information that has less urgency, such as where to send donations, can be displayed on the CCC website or placed in print channels (e.g., newspapers).

People working inside the CCC, and those partnering with it, must never fall into the trap of relying too much on the media. Although the media can serve as an important conduit between CCC leadership and the public (Scholl et al., 2005), it would be a mistake to count on the media to serve the CCCs' or stakeholders' best interests. One way to avoid this mistake is to make sure that information is communicated as accurately and in as timely a manner as possible to the local media outlets. For example, if journalists are asking the spokesperson of an agricultural business about the nature of an accident that happened on-site, it would be important to provide this information as soon as it is verified. On the contrary, if the business does not answer questions related to the accident in a timely fashion, journalists might go to other sources (e.g., former employees) for information. Such sources might not be accurate and could end up misinforming the public. Moreover, law enforcement and elected officials might make decisions on the faulty information, all without the expertise and insight of the organization involved.

Moreover, the media alone are not sufficient to communicate to the public (Scholl et al., 2005). Elected officials and other key decision makers have to make decisions based on the information they have, and if all the insight they have comes from the media, they could end up making decisions that are inappropriate or harmful. On the other hand, communicating directly with the city council or with other municipal leaders before, during, and after events happen can increase the likelihood that decisions are made for the benefit of the community and for the organizations and agencies affected by the crisis. If the CCC is part of this exchange of information, this type of cooperation can also ensure that the CCC remains a vital component of local crisis management efforts, which will enhance further its visibility and credibility in the community.

Because some crisis-related information can be sensitive and carry serious implications, working with the media requires that spokespersons and organizations establish certain rules and protocols. For example, an organization might require media personnel to be escorted during the crisis when they are on the premises. Also, there might be certain staff or personnel who are discouraged from talking to the media. It is advised that all media communication be restricted to the spokesperson(s) or the CEO/head of the organization affected. These and other rules must be clearly established, and the CCC can provide guidance on crafting, communicating, and enforcing such rules. Regardless of the restrictions on who talks to the media, reporters have the right to interview anyone; therefore, these guidelines should be made available to all members of the organization.

The Post-crisis Phase

During the post-crisis phase, the CCC and its partnering organizations need to continue to address the concerns of the public and the stakeholders, as well as monitor and evaluate the effectiveness of existing crisis communication plans (Gonzales-Herrero & Pratt, 1996). Once the major aftershocks have dissipated, the CCC needs to work with affected organizations to continue to manage their image and conduct the necessary damage control. Heath and Millar (2004) point out that the post-crisis phase is an opportunity to examine strengths and weaknesses and to make the necessary changes to the existing plan to prevent repeating the same mistakes. More importantly, continued examination of the current plan might even alert people to signals that could help them avoid a similar crisis in the future.

Audience Issues

Even if the primary impact of the crisis has passed, its effects will likely continue to be felt for an indefinite amount of time. Internal and external CCC partners must continue to monitor current symptoms or impacts as well as be vigilant for those that have not yet emerged. Moreover, message disseminators have to consider the possibility of disaster syndrome (Perry & Lindell, 2003), which is characterized by momentary shock that decreases people's ability to respond appropriately to environmental cues. Other characteristics include passivity and disoriented thinking. The continuation of such symptoms can delay the recovery of the community, hinder efforts to disseminate information properly, and prevent relief and assistance from getting to those who need it.

In response to appropriately crafted and delivered messages, most citizens in shock can still respond effectively and follow instructions without too much difficulty. Effective messaging continues to be crucial and should include three to four core components that support the main message conveyed during the crisis phase (Scholl et al., 2005). Social media posts should be consistent with the main message and continue to link the public to the CCC website or to other credible sources of information, such as government agencies. Timing is important so as not to overwhelm or to try the public's patience (Crelinsten, 1994). If too much time elapses between messages, rumors can surface, thus undermining the efforts of the CCC to keep the community properly informed. Disseminating messages at regularly timed intervals can keep the audience in touch with the event, prevent rumors from spreading, and make the message more meaningful to those who receive it.

Relationship with the Media

The public's understanding of a crisis and its causes could have an impact on people's ability to cope with the aftermath and prevent future threats (Scholl et al., 2005). The public's perception of a crisis depends greatly on the media's portrayal of it, which is why crisis communicators need to work strategically to make sure they continue to maintain a good working relationship with the media. Unfortunately, some news stories could cause adverse effects, such as increased stress and psychological problems (Becker & Thompson, 2005). The CCC network needs to have experts on hand to alter messages as needed to address these problems and, more importantly, to monitor the environment and to adjust communication accordingly to prevent psychological problems from surfacing.

Because media outlets serve multiple audiences (Scholl et al., 2005), the CCC network should be monitoring their sectors, industries, and communities for any adverse effects that might be unique to different segments of the community. Not all segments of a community recover from a crisis in the same way or at the same rate. For example, residents in a neighborhood close to a manufacturing plant might be drinking water that has been contaminated as a result of a disaster. The spokespersons from that plant, the liaisons from related sectors, and other experts in environmental toxicology, health, and municipal services need to work together to make sure the message conveyed to the media is consistent and accurate and does not give the impression that the crisis is completely over for the whole community. In times like this, it is crucial for the media to address the aftermath so as not to leave the community uninformed or to neglect segments of the community who are still experiencing harmful effects.

Post-crisis Assessment

One of the most important post-crisis management processes is conducting an assessment of the actions taken by the CCC network before and during the crisis. It is not enough to formulate a crisis plan and to put it into effect. The outcomes of that plan need to be assessed so changes can be made in anticipation of a future crisis. As part of the centralizing hub of the CCC network, internal CCC personnel should organize and facilitate the outcome assessment. Although members of the affected organizations can participate in the process, it is best to have the CCC personnel who do not have personal or professional interests in the affected organizations lead the assessment effort. This arrangement will help avoid any conflict of interest.

Among other things, the assessment should evaluate the cultural competence of members of the CCC network (Rosen, Greene, Young, & Norris, 2010), timeliness of the messages conveyed to the public (Scholl et al., 2005), consistency across all message forms, technological issues related to web and social media sites, handling of the relationship with the media, coordination among and leadership structure of the CCC network, and containment of rumors and misinformation. These elements can be part of an overall assessment strategy that evaluates three main issues: preparedness, review and monitoring efforts, and training (Institute of Development Studies, 2014). Assessing preparedness involves determining whether relevant organizations have an adequate crisis plan in place, which is consistent with the recommendations put forth by the CCC. Review and monitoring pertains to the existence of efforts to evaluate and update the crisis plan as needed, as well as to document actions that either represent best practices or those that should be avoided. Finally, the assessment should look at the amount and type of training provided, as well as who participated in and successfully completed the training. Training should include information dissemination and skills-based simulation exercises to ensure mastery of tasks.

Conclusion

The purpose of this chapter has been to outline the applied implications of a CCC and provide guidelines for CCC personnel and partners. Community sectors and the organizations and agencies within them have unique needs when it comes to preparing for, and minimizing the effects of, crisis events. More importantly, the CCC can be established to provide a hub or centralizing point around which these sectors and organizations coordinate to create and deliver the most effective crisis communication plans that suit their own needs.

In this chapter, the practical and applied implications of the CCC were covered according to the three major phases of crisis—pre-crisis, during the crisis, and post-crisis. Discussed in each of these phases were the needs of common community sectors, particularly those related to coordination, media relations, audience analysis, and message construction. Also, the CCC network of personnel introduced in Chapter 9 was reviewed and its functions applied to the three crisis phases.

In presenting the research and theoretical premise underlying the book, and in providing guidance and a framework for the CCC, this book serves as a resource for community leaders, decision makers, elected officials, and emergency responders. It also provides scholars directions for future research, such as the comparison of specific crisis prevention strategies across similar communities, and in developing innovative ways to collect and warehouse large amounts of crisis data. Throughout this book, the AMCM has been applied in a justification for, and description for, a CCC. In cooperation with various community sectors, community members can take practical steps during all three phases of crisis to forecast events and to formulate effective crisis communication plans. If individuals are trained at the local level to prepare in advance of such crises, more lives could be saved with more natural ecosystems preserved, resulting in fewer health hazards.

References

American Psychological Association. (2014). How to work with the media: Interview preparation for the psychologist. https://apa.org/pubs/authors/media/index.aspx

Argenti, P. (2002). Crisis communication: Lessons from 9/11. *Harvard Business Review*, 80(2), 103–9.

Becker, J. A. H., & Thompson, S. (2005). The intersection of terrorism, interpersonal communication, and health. In H. D. O'Hair, R. L. Heath, & G. R. Ledlow (Eds.), *Community preparedness and response to terrorism*, Vol. 3: *Communication and the media* (pp. 47–64). Westport, CT: Praeger.

Bowen, S. A. (2012). The ethical challenges of pre-crisis communication. In B. A. Olaniran, D. E. Williams, & W. T. Coombs (Eds.), *Pre-crisis planning, communication, and management: Preparing for the inevitable* (pp. 57–78). New York: Peter Lang.

Crelinsten, R. D. (1994). The impact of television on terrorism and crisis situations: Implications for public policy. *Journal of Contingencies in Crisis Management*, 2(2), 61–72.

Ginzel, L. E., Kramer, R. M., & Sutton, R. I. (1993). Organizational impression management as a reciprocal influence process: The neglected role of the organizational audience. *Research in Organizational Behavior*, 15, 227–66.

Glynn, M. A. (2000). When cymbals become symbols: Conflict over organizational identity within a symphony orchestra. *Organization Science*, 11(3), 285–98.

Gonzales-Herrero, A., & Pratt, C. B. (1996). An integrated symmetrical model for crisis-communications management. *Journal of Public Relations Research*, 8(2), 79–105.

Hayes, D., & Patton, M. (2001). Proactive crisis-management strategies and the archaeological heritage. *International Journal of Heritage Studies*, 7(1), 37–58.

Heath, R. L., & Millar, D. P. (2004). A rhetorical approach to crisis communication: Management, communication processes, and strategic responses. In D. P. Millar & R. L. Heath (Eds.), *Responding to crisis: A rhetorical approach to crisis communication* (pp. 1–17). Mahwah, NJ: Lawrence Erlbaum.

Institute of Development Studies. (2014). TSM assessment protocol: A tool for assessing crisis management planning performance. Eldis. http://www.eldis.org/go/home&id=39526&type=Document#.UwGLKR0o7mI

Jaques, T. (2012). Issue management as a strategic aspect of crisis prevention. In B. A. Olaniran, D. E. Williams, & W. T. Coombs (Eds.), *Pre-crisis planning, communication, and management: Preparing for the inevitable* (pp. 17–35). New York: Peter Lang.

King, G. I. (2002). Crisis management and team effectiveness: A closer examination. *Journal of Business Ethics, 41*, 235–49.

Massey, J. E. (2001). Managing organizational legitimacy: Communication strategies for organizations in crisis. *Journal of Business Communication, 38*(2), 153–70.

Millar, D. P. (2004). Exposing the errors: An examination of the nature of organizational crises. In D. P. Millar & R. L. Heath (Eds.), *Responding to crisis: A rhetorical approach to crisis communication* (pp. 19–32). Mahwah, NJ: Lawrence Erlbaum.

Mitroff, I., Alpaslan, M. C., & Green, S. E. (2004). Crises as ill-structured messes. *International Studies Review, 6*(1), 165–82.

Moore, S. (2004). Disaster's future: The prospects for corporate crisis management and communication. *Business Horizons, 47*(1), 29–36.

Olaniran, B. A., Scholl, J. C., Williams, D. E., & Boyer, L. (2012). Johnson and Johnson phantom recall: A fall from grace or a re-visit of the ghost of the past. *Public Relations Review*, 38, 153–55, doi:10.1016/j.pubrev.2011.08.001.

Pauchant, T. C., & Mitroff, I. I. (1992). *Transforming the crisis-prone organization: Preventing individual, organizational and environmental tragedies*. San Francisco: Jossey-Bass.

Pearson, C. M., & Clair, J. A. (1998). Reframing crisis management. *Academy of Management Review, 23*(1), 58–76.

Perry, R.W., & Lindell, M. K. (2003). Preparedness for emergency response: Guidelines for the emergency planning process. *Disasters, 27*(4), 336–50.

Polevoi, L. (6 September 2013). Tips for being interviewed by the media. Intuit Small Business Blog. Retrieved from http://blog.intuit.com/marketing/tips-for-being-interviewed-by-the-media/

Rosen, C. S., Greene, C. J., Young, H. E., & Norris, F. H. (2010). Tailoring mental health disaster services to diverse needs. An analysis of 36 crisis counseling projects. *Health and Social Work, 35*(3), 211–19.

Scholl, J. C., Williams, D. E., & Olaniran, B. A. (2005). Preparing for terrorism: A rationale for the Crisis Communication Center. In D. O'Hair, R. Heath, & G. Ledlow (Eds.), *Community preparedness and response to terrorism*, Vol. 3: *Communication systems and the media's role in community service*. Santa Barbara, CA: Greenwood Publishing Group.

Seeger, M. W. (2002). Chaos and crisis: Propositions for a general theory of crisis communication. *Public Relations Review, 28*, 329–37.

Seeger, M. W. (2006). Best practices is crisis communication: An expert panel process. *Journal of Applied Communication Research, 34*(3), 232–44.

Sterling, G. (2015, May 5). It's official: Google says more searches now on mobile than on desktop. Search Engine Land. Retrieved February 3, 2015, from http://searchengineland.com/its-official-google-says-more-searches-now-on-mobile-than-on-desktop-220369

WebMD. (2014). Meningitis: A topic overview. http://www.webmd.com/children/vaccines/tc/meningitis-topic-overview

Case Study

Meningitis Strikes a Community

Issues management: What follows is an illustration of a crisis management situation to which the functions of the CCC network could be applied. To show how the CCC operates within the community and collaborates with its external partners, a hypothetical but realistic situation will be discussed in relation to the three phases of the crisis. The case will be described throughout each phase as it is addressed and discussed by the CCC network. The case study will illustrate how a CCC can supplement existing efforts at crisis anticipation and management.

Unfortunately, many businesses and organizations have crisis plans that are ill-structured and yet complex (Mitroff, Alpaslan, & Green, 2004), which often leads to the mishandling of crises when they occur and the inability to detect important signals. While many crises are avoidable, it is easy for businesses and noncommercial sectors either to miss the signals of potential crises or not to take timely action because of individual carelessness, system breakdowns, and inefficient practices (Argenti, 2002; King, 2002). Such is likely the case with the emergence and spread of meningitis, an illness that is passed through sneezing, coughing, or close contact (WebMD, 2014). The thought of meningitis likely raises fear among many people, who might be misinformed about its causes, symptoms, and outcomes. Such an outbreak might be one of potential problems raised during one or more issues management discussions among sector liaisons, whether or not it has even occurred.

As part of the issues management process, liaisons and spokespersons should identify and discuss the public health threats relevant to their community, even if a particular outbreak is not detected. As part of anticipatory planning, liaisons from the private sector might reveal the health codes and regulations they have to follow, the procedures that outline steps to take in the event of an outbreak, and the methods already in place to inform the public. Liaisons from public institutions, such as universities, might also be able to bring to light previous cases. Public relations experts and health communication scholars can explore the best methods for disseminating information. Those in the medical sector would be able to inform others in the CCC network of the difference between viral (more common but not usually life-threatening) and bacterial meningitis (less common, but serious) and common symptoms, which include headaches, fever, seizures, vomiting, stiff and painful neck, and difficulty staying awake. Compiling such information and formulating communication plans in advance will enable the CCC to work with the appropriate sectors and organizations in adapting a message for the public and be able to take timely action when necessary. Such an issues management process is more effective if liaisons and spokespersons meet regularly to scan the local environment and to discuss the signals they detect. These signals can be incorporated with other information about the issue into a digital asset management system (see Chapter 9), which can be accessed by those in the CCC network who have communication-related responsibilities.

If a noticeable increase in meningitis cases is detected by one or more spokespersons or liaisons, but the local media hasn't yet picked up the story, it could be that the threat has not reached crisis levels. In this case, the CCC network can prepare messages and confirm that plans are in place in the event of a community-wide epidemic. The CCC can be on standby to deal with the media and coach spokespersons who might be called upon to address the public. Finally, CCC staff can prepare the facility for any meetings that would need to occur during an actual crisis and have on hand materials to disseminate to the media and to the public.

The crisis unfolds: The effects of crises can be overwhelming and threaten a community's financial security, public safety, and environmental stability (Hayes & Patton, 2001; King, 2002; Pearson & Clair, 1998). This is especially true of a meningitis outbreak. For one thing, such a crisis would threaten the bottom line of the restaurant rumored to be the source of the outbreak, overwhelm local medical facilities, and even undermine the authority of local officials, who face an angry and frightened community. As the threat of meningitis is reclassified as a full-blown outbreak, medical experts now verify that normal levels of infection have surpassed acceptable thresholds, and the media are clamoring to talk to anyone who will shed light on the situation.

At the onset of the outbreak, the CCC network needs to act quickly to share information among its members, come to an agreement as to the nature and causes of the outbreak, generate a core message, provide relief and assistance to community members in need, and formulate a crisis management plan that minimizes the impact of the crisis. During the process of verifying the nature and cause of the crisis, a spokesperson from the local hospital might alert the medical sector liaison to a group of patients who present identical symptoms of bacterial meningitis, and that most of these patients had eaten at a local restaurant chain within the past week. The liaison representing the food industry could confirm this report, and a plan could be put in place to isolate the disease and provide medical treatment to those affected, as well as treat those who were put at risk.

This kind of crisis has financial implications, not only for the restaurant chain at the center of the disaster, but also for some of those exposed to infection who have limited access to health care. The board of directors, executive director, liaisons, and spokespersons working in the CCC network need to be able to assess the financial impact of the outbreak, estimate how much treatment will cost, make decisions as to the availability of medications, determine who in the community is required to receive treatment, and establish criteria for financial assistance or reimbursement for medical costs. For instance, liaisons and spokespersons from other sectors (e.g., banks and grocery stores) might be contacted to see if they can provide monetary assistance or supplies to those who were exposed to the infection but face financial obstacles to getting treatment.

The CCC network also has to deal with the health and safety implications of a meningitis outbreak. Segments of the public might be misinformed about the nature of the infection, what causes it, and how it is spread. Misinformation and information overload could lead to panic, disassociation, or apathy. Medical personnel need to confirm the nature and type of infection, and decide how much information the audience needs to know. This is where the delicate balance comes in regarding the relationship with the media. The medical liaisons and spokespersons might decide who should talk to the press, what message components should be communicated, and which media channels should be used. Spokespersons or liaisons in the medical community need to work closely with CCC internal staff to ensure that the media are dealt with efficiently, and that all information is accurate and consistent across the board.

While some urgent information is chosen for immediate release on television and Internet live streaming, other pieces of information can be generated and revised for incorporation onto the CCC website to allow the public to access additional information at their convenience. A communication review of print and electronic materials is recommended to ensure accuracy and consistency across all media channels used. In addition, information that has already been stored on a digital asset management system (e.g., common symptoms, treatments, incubation periods, frequently asked questions, and common myths) can be retrieved in a short amount of time and incorporated into a message template.

Learning from the crisis: The meningitis case study during the post-crisis stage addresses the sustained monitoring of the environment and continued dissemination of information. In the post-crisis assessment, the CCC network, in conjunction with the board of directors and internal staff, can examine the timing of the messages that went out to the media and the public. In their monitoring of public perception and response to the crisis, it might be determined that, despite the CCC's best efforts, the public was not properly informed, and even contributed to the spread of false rumors about the meningitis outbreak. It might be determined that too much time passed between news releases. In the future, the plan will involve news releases and press conferences that are closer together to minimize instances of misinformation.

Another element of assessment involves the accuracy of the information given to the public. For instance, medical experts can monitor the press conference given by the restaurant spokesperson and provide this person feedback about her or his effectiveness in communicating health-related information. Another assessment element might involve the cultural competence of the CCC staff taking inquiries from the public. For instance, it turns out that the primary language of several people who called the CCC hotline was not English. Having not prepared for this possibility, the CCC executive director did not think to recruit some bilingual volunteers. In making recommendations for future recruiting efforts, the CCC liaison suggests that the CCC has on its volunteer roster, a few people who speak languages other than English, and that at least one of these individuals should be at the phones during times of crisis to take inquiries.

Discussion Questions:

1. Which sectors within a community might be most affected by the threat of a meningitis outbreak, and why?
2. What kinds of messages should a CCC help put in place to prepare for a meningitis or other type of health threat to the public?
3. What are the lessons learned based on the limitations pointed out in this case study? How would such lessons be used to prepare for future crisis events?

INDEX

B

C

D

E

F

G

H

I

J

K

L

M

N

O

P

Q

R

S

T

U

V

W

Y

Z

www.ingramcontent.com/pod-product-compliance
Ingram Content Group UK Ltd.
Pitfield, Milton Keynes, MK11 3LW, UK
UKHW021852100926
8556IPUK00020B/1190

9 781433 124334